Political Street Art

Recent global events, including the 'Arab Spring' uprisings, Occupy movements and anti-austerity protests across Europe have renewed scholarly and public interest in collective action, protest strategies and activist subcultures. We know that social movements do not just contest and politicise culture, they create it too. However, scholars working within international politics and social movement studies have been relatively inattentive to the manifold political mediations of graffiti, muralism, street performance and other street art forms.

Against this backdrop, this book explores the evolving political role of street art in Latin America during the twentieth and early twenty-first centuries. It examines the use, appropriation and reconfiguration of public spaces and political opportunities through street art forms, drawing on empirical work undertaken in Brazil, Bolivia and Argentina. Bringing together a range of insights from social movement studies, aesthetics and anthropology, the book highlights some of the difficulties in theorising and understanding the complex interplay between art and political practice. It seeks to explore 'what art can do' in protest, and in so doing, aims to provide a useful point of reference for students and scholars interested in political communication, culture and resistance.

It will be of interest to students and scholars working in politics, international relations, political and cultural geography, Latin American studies, art, sociology and anthropology.

Holly Eva Ryan is Lecturer in International Political Sociology at Queen Mary University of London, UK.

Routledge Research in Place, Space and Politics
Series Edited by Professor Clive Barnett, Professor of Geography and Social Theory, University of Exeter, UK.

This series offers a forum for original and innovative research that explores the changing geographies of political life. The series engages with a series of key debates about innovative political forms and addresses key concepts of political analysis such as scale, territory and public space. It brings into focus emerging interdisciplinary conversations about the spaces through which power is exercised, legitimized and contested. Titles within the series range from empirical investigations to theoretical engagements and authors comprise of scholars working in overlapping fields including political geography, political theory, development studies, political sociology, international relations and urban politics.

Political Street Art

Communication, culture and resistance in Latin America

Holly Eva Ryan

LONDON AND NEW YORK

First published in paperback 2019

First published 2017
by Routledge
2 Park Square, Milton Park, Abingdon, Oxon OX14 4RN

and by Routledge
711 Third Avenue, New York, NY 10017

Routledge is an imprint of the Taylor & Francis Group, an informa business

British Library Cataloguing-in-Publication Data
A catalogue record for this book is available from the British Library

Library of Congress Cataloging-in-Publication Data
Names: Ryan, Holly Eva, author.
Title: Political street art : communication, culture and resistance in Latin America / Holly Eva Ryan.
Description: Abingdon, Oxon ; New York, NY : Routledge is an imprint of the Taylor & Francis Group, an Informa Business, [2017] | Series: Routledge research in place, space and politics
Identifiers: LCCN 2016030843 | ISBN 9781138852884 (hbk) | ISBN 9781315723211 (ebk)
Subjects: LCSH: Street art—Political aspects—Latin America. | Public spaces—Political aspects—Latin Ameria. | Protest movements—Latin America. | Latin America—Politics and government—1980–
Classification: LCC ND2602 .R83 2017 | DDC 306.4/7098—dc23
LC record available at https://lccn.loc.gov/2016030843

ISBN: 978-1-138-85288-4 (hbk)
ISBN: 978-1-138-38492-7 (pbk)
ISBN: 978-1-315-72321-1 (ebk)

Typeset in Times New Roman
by Apex CoVantage, LLC

Contents

Figures

Acknowledgements

To complete this book, I relied on the support, encouragement, and generosity of many individuals and institutions. Thomas Davies, James Dunkerley, Roland Bleiker, Jelena Obradovic-Wochnik, Sophie Harman and many other academic colleagues offered constructive critique and indispensable advice for which I am extremely grateful. The Department of International Politics at City University London made this publication possible by funding my doctoral research and providing a supportive environment for me as a PhD student and later as an Honorary Visiting Research Fellow. Staff at ESCALA (University of Essex) helped tremendously by allowing me to trawl through their archives for hours on end. Additionally, the Crick Centre and Department of Politics at the University of Sheffield, where I worked as Postdoctoral Researcher whilst preparing the manuscript, offered a stimulating intellectual home where interdisciplinary work was positively encouraged.

The greatest credit is due to the inspiring art-activists whose stories I have touched upon in this book. It has been my honour to have met a great many courageous and interesting people during research trips to Latin America. Words cannot express how appreciative I am for their time, openness and trust. I am especially grateful to: Monica Hasenberg, Laura Salazar de la Torre, Graciela Carnevale, Paulo Ito and Stanislaw Czaplicki Cabezas for making their photographs available for use in this book. A huge thanks also goes to Jaime, Neta, Lina and Anita Prades for their tremendous hospitality and for giving me access to their wonderful archive of slides.

Thanks to the editing team at Routledge – particularly Faye and Priscilla – for their support and guidance during the publication process. I am also grateful to the series editor, Clive Barnett for all his efforts in bringing together the Routledge Research in Place, Space and Politics series.

Final thanks are due to my nearest and dearest: my parents Runeika Ryan and Edward (Ted) Ryan who brought me up to think critically and have always supported me; and, Thomas Orrell, my life partner who has proofread countless drafts and reminds me to take days off every now and then.

1 Introduction

Latin America is a large and complex region featuring diverse cultures, landscapes, habitats and histories. In his memoirs, the Chilean poet Pablo Neruda described Latin America as the 'continent of hope', alluding to both the unfulfilled promise of the region's peoples and their unwavering spirit of perseverance in the face of social and political upheaval. Although Latin American states have made enormous political and economic strides in recent decades – from the consolidation of democratic governments and improvements in the status and recognition of indigenous peoples to reductions in inequality – there remain a great many challenges to freedom and prosperity. Moreover, the region as a whole remains decisively shaped by dramatic social and political change seen in the last century. Over the course of the twentieth century, Latin America bore witness to swings back and forth between civilian and military rule, interventions and alliances formed on the basis of superpower rivalry during the Cold War, as well as ongoing struggles to come to terms with the social and economic fallout from the colonial era. Together, these experiences have made Latin America an especially fertile ground for scholarly investigations into questions of power and 'politics by other means'.

A range of theorists, including Oliver et al. (2003), Goodwin and Jasper (2014) and Ramos and Rodgers (2015), have described social movement activity as 'politics by other means'. In so doing, they appropriate Carl von Clausewitz's description of war to emphasise how the aspirations, methods and consequences of political activism represent an extension of conventional institutional and democratic politics. Of course, social movements, defined as organised sets of constituents pursuing a common political agenda of change over time (Batliwala 2008), do interact with state systems to produce modest reforms such as enhanced political participation and policy adjustments. However, their actions can also contribute to much broader processes of socio-political change, even wholesale shifts in the political system. Social movement actors have often found novel ways to circumvent, bypass or disrupt the system and their influence may even extend to sparking insurrection and revolution (Johnston 2011).

To depict forms of political activism as 'politics by other means' is to propose wide and encompassing parameters for what constitutes politics and who counts as a relevant political actor. For Rancière (2010; 2011), what is recognised as political action is far from fixed, and this dynamism is built into the very meaning

Figure 1.1   Latin America

of 'doing politics'. He writes, '[p]olitical action consists in showing as political what was viewed as "social", "economic" or "domestic". It consists in blurring the boundaries'. More to the point, social movements interact with state systems and societies in a wide variety of ways, often 'blurring boundaries' by recasting economic, social or cultural issues as political ones and bringing them to the attention of new constituencies. In recent decades, social movement scholars have put a great deal of energy into understanding how activists mobilise support and articulate their concerns to broader audiences (see for example: Snow et al. 1986; Snow and Benford 1988; Taylor and Van Dyke 2004). But activists don't just contest and politicise culture, they create it too. In Latin America, cultural producers – poets, painters, writers and musicians – have often been at the forefront of movements for political change. However, existing social movement scholarship offers only a very limited toolkit for understanding the role of art and creative expression in contentious politics. Recent publications on Latin American social movements in particular tend to neglect these influential modes of political claim-making and expression, focusing more on the evolving political environment and the complex relationship between neoliberalism, globalisation and democracy in the post-Cold War era (see Johnston and Almeida 2006; Petras and Veltmeyer 2011). The research presented in this book seeks to address these gaps by exploring the cultural productions of activists in Latin America and by honing in on a specific type or category of 'politics by other means'. This book is about 'political street art'.

The category of 'political street art'

Most people would claim to have a pretty good idea of what street art is, and yet many of the most common definitions, on close inspection, appear somewhat strained or inadequate. As a first caveat, the characterisations that are offered of street art are often bound up with individuals' particular and prior experiences, ideologies and pre-formed aesthetic preferences. So for some, street art may be seen as a symbolic re-appropriation or 'taking back' of the public space – a democratising act with anti-capitalist and/or anti-authoritarian undertones. Meanwhile for others, inscriptions on urban surfaces remain an act of nuisance or vandalism: something to be penalised and discouraged rather than conflated with 'art' as such. Of course, as Belfiore and Bennett (2008) highlight, the question of what constitutes 'art' in the first place has puzzled theorists since the very beginnings of philosophical enquiry and debates are still ongoing. Today, the main battle lines are drawn between those who ascribe to a *functionalist* understanding – that 'art' must serve some preordained purpose, whether spiritual, pedagogical, cultural or political – versus those who ascribe to a procedural or *institutionalist* understanding. The latter suggest that only the aggregation of institutions involved in the 'artworld(s)' such as museums, galleries and critics may confer the label of 'art' unto an object (ibid.). Depending on which understanding one ascribes to, street art is likely to be recognised and valued in markedly different ways. Although certain forms of street art are slowly gaining institutional acceptance, the prefix 'political' quite evidently implies that the street art under focus in this book is understood to serve a purpose.

Existing work on political street art tends to focus on one of three characteristics: its social nature; ephemerality; and, its relationship to the public space. Robert Sommer's (1975) exploration of muralism in the slums and ghettos of North America provides one of the earliest academic engagements with street art. For him, street art is a complex phenomenon 'that includes many ingredients – painting done in the presence of an audience, the rhythm of crowds, interaction with local gangs, the hostility of drunks, the watchfulness of the police, zoning regulations and sign ordinances, and protection and maintenance by disinterested third parties' (Sommer 1975: 7). Sommer's emphasis on interaction with audiences – gangs and drunks – as a part of the street art production process underlines the *social nature* of the phenomenon. Alongside Sommer's definition, Lyman Chaffee's (1993) seminal study of *Political Protest and Street Art* makes reference to collective expressions through largely ephemeral media including posters, wall-painting, murals and graffiti. Notably, Chaffee also considers 'auxiliary forms' of ephemera – for example political stickers, t-shirts and banners. He treats these as individual expressions, although they are nonetheless social in the sense that they tend to demonstrate support for a shared cause or system of values.

The *ephemeral quality* of street art is another characteristic brought into focus by muralists, graffiti writers and the various commentators that seek to define their activities. Chicago muralist John Weber wrote in his *Technical Notes on Materials and Techniques for Mural Painting* (1972) that even under the very best of conditions, an exterior mural would be unlikely to survive for more than thirty years. Chaffee highlights that other forms – graffiti, posters and street performance, for example – tend to have a much shorter 'public' life, being routinely battered by the elements, painted over or otherwise destroyed. Yet, for those that produce street art, the issue of ephemerality and the fact that works may not survive long is rarely of great concern. Indeed, many street artists attribute a kind of democratic value to the ongoing cycle of creation and destruction. As London street artist Mutiny (2015) writes,

> [Having your work painted over] is all the natural part of the cycle of street art. All street artists know this and accept it as part of the deal. This is, after all, the ephemeral nature of street art. It's mine and it's yours – it belongs to everyone.

As such, street art often appears to have the distinguishing feature of moving along with the times. This stands in contrast to the aims of museums and the more traditional galleries, in which temperature, light, humidity and interactions with artworks are carefully managed so that artifacts might be preserved for the long term.

Linked to claims about the social nature and ephemerality of street art are a number of questions about its relationship with and to *the public space*. In the quotation from Sommer (1975) above, the category of 'street art' covers that which is produced outdoors, where the participation, reactions and rhythms of the community may rub off on the artist/s and find their way into the meaning of the work. For Chaffee (1993) and others, the public location and visibility of street art interventions are key to their political meaning. By disturbing or subverting the aesthetic and cultural codes established by governing bodies and corporate advertisers, street art provides a 'talk back' mechanism for the public. It may even offer a cue for witnesses to envision the world around them differently.

Yet, the relationship between street art and the public space may be rather more complex than this since there are, as Riggle (2010) notes, innumerable ways of using the street as an artistic resource. One can make use of open spaces for urban displays and installations, incorporate objects found on the street into a 3D constructions, or even 'jam' existing technologies of communication that have been employed on behalf of states and corporations. Notably for Riggle (2010), street art is distinguishable as such, only if: i) creative and material use of the street is internal to the meaning of the art; and, ii) if that meaning is fluid and free from functional constraints imposed by the market. However, the extent to which street art can free itself from market imperatives is today in question. As technological advances have made it simpler to document, map and share images across the world, the emerging popularity of *certain forms* of street art has not gone unnoticed by governments and corporations who have themselves taken steps to incorporate the medium into urban regeneration programmes, advertising and commerce. Since the 2000s, some governments have moved to decriminalise graffiti and others have created 'free painting zones' in urban centres. Works by internationally recognised street artists now sell for hundreds of thousands of pounds on gallery circuits. In 2012, as London councils began a street 'clean-up' operation ahead of the Olympic Games, street art interventions by Britain's Banksy were granted exceptional status (Café 2012). These shifts prompt us to ask political questions: why, for example, should states seek to neutralise and appropriate street art in the first place? And importantly, to what extent does the increasing marketisation of street art undermine its power as a disruptive technology and form of critique?

With the above-mentioned debates and sensitivities in mind, this book offers up the term 'political street art' as a loose category for interventions whose creative and material use of the street is in some way tied to their political meaning. The definition is deliberately broad and seeks to make space for the consideration of overt and non-overt forms of politics that manifest *in and around* street art. In other words, it holds that to be *political* is not just to express political opinions but rather to be oriented toward society and to engage with its variegated terrains of power.

Researching political street art in Latin America: choices and challenges

Political street art – posters, wheat-pastes, graffiti, murals and street performance, among other forms – has a familiar, distinctive and indeed pervasive presence in cities across Latin America today. To take one recent example, a week before the inaugural game of the 2014 World Cup in Brazil, a vibrant wall-painting of a boy crying hysterically as he is served up a football instead of dinner, went viral across social media outlets. The painting, by Sao Paulo based artist Paulo Ito, was widely interpreted as an indictment of the *Fédération Internationale de Football Association* (FIFA) and became a symbol of popular discontent with the World Cup (Ryan 2014). Political street art also has a long history in the region, emerging alongside the shift to mass politics to provide a 'low technology medium of communication' and mobilising tool for pro-system and anti-system forces alike (Chaffee 1993).

As this book will endeavour to show, political street art has provided a unique resource for groups that have been denied access to institutionalised channels of communication. Time and again, Latin Americans have taken to the streets and to the walls, armed with paintbrushes, aerosol cans, stencils or posters. They have intervened in the public space in ways that amplify and give form to their opinions

Figure 1.2   'Meme' by Paulo Ito, 2014
(Photo courtesy of Paulo Ito)

and feelings, demonstrating the power of political street art in combatting forms of 'excommunication' (Mattelart 2008) that arise from socio-economic inequality and repression.

While it may be the case that under authoritarian regimes city walls are one of 'the only places where [artists can] talk back to tyrants' (NPR 2013), to suggest that street art is an inherently participatory, non-violent and democratic model of expression would be inaccurate. Indeed, in Latin America, the state has itself been no stranger to the production of street art. Mendoza and Torres (1994) explain that the history of muralism can be linked to the earliest periods of Spanish settlement in the Americas, with the first murals representing the doctrines of both Church and

State. In the more recent past, muralism and other forms of wall-painting have often accompanied state experiments in both revolutionary socialism and authoritarianism.

The legacy of the Mexican muralists in particular, has been felt across the region. The Mexican Mural renaissance was kick-started shortly after the end of the Mexican Revolution in 1921, when José Vasconcelos, the revolutionary government's new Education Minister, publicly urged the country's artists and intellectuals to 'leave [the] ivory towers to sign a pact of alliance with the Revolution' (Benjamin 2010). Vasconcelos set up a major arts programme commissioning local artists to create works for the walls of prominent public buildings in Mexico City. The artists David Siqueiros, Diego Rivera, and José Clemente Orozco – known as *los tres grandes* – led in the production of these murals, which were aimed at communicating Mexico's post-revolutionary ideals and national history to the masses, of whom many were illiterate. Between the 1920s and 1950s, the art-activists cultivated a style of painting that not only sought to re-define Mexican identity in the wake of the Revolution but also had reverberations across the region.

Exposing the history of political expression, communication, and defiance through street art in Latin America is an important corrective to certain distortions that have emerged within the small body of existing scholarly work on the medium. One commonly repeated myth is that street art – and particularly graffiti – began its life in the hip-hop movement of 1960s New York and that 'the new *writing* that appeared on New York City's public walls in the late 1960s and early 1970s … was unprecedented in modern history' (Austin 2010). More recently, as Charles Tripp (2012; 2013) highlights, commentary accompanying the 'revolutionary street art' of the 'Arab Spring' uprisings has inadvertently reinforced the notion that 2011 represents a kind of 'year zero' for political graffiti and other creative forms of protest. These inaccuracies not only neglect the rich and multi-faceted history of political street art in Latin America, they also reveal Anglo-American-centrist and presentist biases in the literature.

Researching political street art in the Latin American context poses several significant challenges. The first is that of representativeness. Latin America encompasses an expanse of countries dominated by two similar languages and sharing interwoven but distinct cultures and histories. The emergence of political street art as a contentious performance in Latin America therefore takes place against a complex backdrop that includes the transfer of people, ideas and images horizontally, across the Atlantic and also vertically, between the Americas (Miller 2006). It includes processes of nation-building that have, at times, involved the importation and emulation of Western political institutions, and at others involved endeavours to subvert, purge and replace European and North American influences. Moreover, although Latin America boasts a diverse citizenry – the region is home to around 550 million people, around 50 million of whom are classed as indigenous and another 130 million as afro-descendants (The Project on Ethnicity and Race in Latin America 2016) – questions about ethnic difference were long suppressed, leaving 'indigenous groups and people of African descent

[…] economically disadvantaged and politically marginalised well into the twentieth century' (Yashar 2015).

Street art is shaped by this complex mosaic but it also mediates within it in important ways. In fact, we can understand street art as an alternative way of 'doing politics'. Specifically, political expression through street art production can be understood as an example of 'everyday politics', or even 'everyday resistance', in that it provides a space and opportunity to contest and question dominant cultural codes and conventions. Indeed, the presence and content of murals, graffiti and street performances can act as an important 'barometer of a community's identification with its history, traditions and … cultural heros' (Mendoza and Torres 1994: 78). It can also act as a cue for rethinking and remaking culture itself. Therefore, although it might seem cumbersome to devote a chapter to each country rather than tackle the various street art interventions by 'theme' or by 'form', this choice reflects a will to situate the various interventions in their political, historical and social context. As far as possible, this approach seeks to avoid flattening out the differences between states, cultures and communities.

However, related to the challenge of representativeness are several issues that arise in the process of working with – that is, documenting and interpreting – political street art. It is worth noting from the outset that this book works from the premise that street art can provide us with a useful source of social and popular history: that we can come to know something about politics and about society through exploring it. Notwithstanding Platonic anxieties about the imitative and flawed nature of art (see *Republic* 3), there is a strong basis for this claim. Working in the context of post-transition Brazil – where much formal documentation of the regime's violence has been lost or destroyed – social historians have led the call for new initiatives to unlock and amplify the forgotten past. While Bickford (1999; 2000) speaks of an 'archival imperative', Serbin (2006) implores fellow scholars to work with marginalised, overlooked, unpopular and even distasteful stories in order to build a fuller picture of the authoritarian experience. Here – and elsewhere – street art, as an ephemeral medium with a 'built-in impermanency' can provide an expedient, if unconventional, form of 'evidence'. Taking the example of political posters, Tschabrun (2003:305) explains:

> The built-in obsolescence of political posters endows them with a quality not always shared by more permanent forms of primary source material, namely that the creators of posters rarely, if ever, thought they were creating historical evidence… political posters may sometimes attest to underreported or even illegal activities that may be documented in no other way.

By extension, it is possible to think about the ways in which other street art forms can provide a more candid insight into public feeling, political opinion and the strength of circulating ideologies. As Chaffee (1993: 4) explains, 'in the unending process of social conflict and state formation, street art [has been] a tool for analyzing and describing that process'.

Yet, the 'built-in obsolescence' or ephemerality of street art is also a source of challenge for researchers. To begin with, street art is hard to reliably document. Countless interventions fade away unrecorded with each year that goes by, meaning that we can only ever work with a small sample of what has been produced. Additionally, street art interventions are quite often deliberately anonymised, making it hard to trace who has produced them, when, and why they have done so. As a result, there will always be interventions, expressions and stories that are missed.

Street art documentation also places a particular set of responsibilities on the researcher. Capturing a mural or piece of graffiti on camera and sharing it online means 'removing it from its space-time frame and giving it a limited immortality' (Sommer 1975). Today, the opportunities for street art documentation have grown thanks to the advent of web 2.0, the birth of social media and new interfaces such as Flickr, Instagram and Google Street Art,[1] but the use of these new media also draws the issue of responsibility into even sharper focus. Researchers ought to reflect on their contributions to the global circulation of images, not in the least because through their processes of selection and dissemination they have the power to shape or limit knowledge and experience as well as affect lives. They can, for example, elevate certain experiences and individuals from relative obscurity to the public interest. Immortalising certain political street art interventions and the artist-activists who produce them may amplify some expressions and claims at the expense of others. It could also draw unwanted attention to those artists who don't wish to be 'outed' for risk to their reputation, legal standing or safety. Meanwhile, removing street art from its space-time frame may also differ starkly from the artists' intensions and desires: street art interventions may well be very intentional, but not actually intended as art.

Intent, or what Shaw (2001: 275) describes as the 'question of aboutness' poses a further challenge for researchers interested in the production and reception of political street art. It may be the case that street artists have a very precise message in mind when painting, cutting stencils or writing graffiti. However, as media theorists, art historians and others have convincingly argued, we can always count on some degree of mismatch between authorial intent as encoded in the art intervention and the reception or decoding of it. According to Stuart Hall's 'Encoding-Decoding' model of communication, senders encode meaning in their messages that accord with their ideals and views but these messages are then decoded by the receivers in ways that align with a different set of ideals and views. This may lead to the receiver understanding something very different from what the sender intended (Hall 1993). Things are further complicated when we allow for the possibility that street art production may be driven by sentiments or even created in the spirit of aesthetic play[2] such that meanings are ambivalent until of course, we pin them down with linguistic categories. The implication of this is that there is always something that remains in excess of our descriptions of art, but this does not invalidate attempts to describe and interpret images. Indeed, W.J.T. Mitchell recasts the question of intent from the author to the image itself, arguing that, 'What pictures want is not the same as the messages they communicate or the effect they produce; its not even the same as what they say they want... pictures don't know what they want; they *have* to be helped to recollect it through dialogue with others' (Mitchell 1996: 81, italics added).

To help 'make sense' of the pictures and interventions explored within this text the street art practitioners themselves have, where possible, been invited to engage in such reflection and dialogue. In their descriptions these groups and individuals shed light on the full register of elements and drives at play in their street art. In so doing they bring us somewhat closer to an engagement with the 'sticky entanglements of substances and feelings, of matter and affect' (Highmore 2010: 139) that facilitate, motivate and complicate political action; and that are central to the ways in which street art production works over, with and against the status quo.

For researchers most concerned with building a robust scientific model that produces certainty and generalizable results, the study of political street art is not going to deliver. However, to those more open to an interpretive quest for knowledge and to embarking on research that carries a degree of risk, working with street art and those who produce it can be extremely rewarding. There is strong evidence to suggest that street art can act as a source and conduit of political power. Indeed, '[i]f groups in democratic and authoritarian systems utilise political art, they must believe it serves a useful function and produces an impact' (Chaffee 1993: 24). But what is the nature of this impact? How does street art play a political role? And how does this role shift and evolve over time?

This book, based on research undertaken in three quite different Latin American states – Brazil, Bolivia and Argentina – takes the first steps in building a case and framework for examining political street art as a 'contentious performance', adapting from Tilly (2008). Building bridges between literatures on social movements, aesthetics and anthropology, it aims to provide an in-depth exploration of the role of street art in political protest, expression and claim-making.

The structure of this book

Building on from this introduction, Chapter 2 sets the context for this study by exploring forms of 'excommunication' in Latin America. Using the term as deployed by Armand Mattelart, the chapter outlines two distinct patterns or characteristics of Latin American politics that have separated persons from processes, channels and circuits of informational and material exchange. These are poverty and inequality on the one hand and, authoritarian excess on the other. The chapter then goes on to build the case for approaching street art as a unique category of contentious performance: one that requires both an instrumental and an aesthetic understanding. In so doing, the chapter underlines ways in which street art can express and foster resistance to forms of 'excommunication'. It advances a conceptual toolkit for the study of this particular medium or brand of activist-art and in so doing calls for an 'aesthetic turn' in social movement studies.

Chapters 3, 4 and 5 offer exploratory case studies from three different Latin American states, drawing upon archival research, observation and unstructured interviews carried out with current and former street art practitioners, activists, curators and academics between the years 2011 and 2015. Chapter 3 utilises Lyman Chaffee's account of 'Political Protest and Street Art' as a springboard for exploring the history and politics of Brazilian street art through periods of authoritarian and civilian rule. It builds on Chaffee's seminal text by providing new examples and analysis that illuminate the varied and important role that street art has played in Brazilian politics. The first half of the chapter addresses questions such as 'what are the challenges to studying Brazilian street art?' and 'what makes street art posters impactful?', while focusing on early examples of street art. The second half of the chapter traces the emergence and evolving practice of *Grupo Tupinãodá* in particular. *Grupo Tupinãodá* can be considered Sao Paulo's very first street art collective. The group came together during the authoritarian period, developing novel creative and expressive forms to intercede and critique the circumscription of political rights, freedoms and brutality of the time.

Chapter 4 charts the development of political street art in Bolivia: a story inseparable from the long and arduous struggle for substantive social and political equality for the country's indigenous population. It offers a journey through a long twentieth century, exploring the ways in which political street art developed up until the time of the 1952 National Revolution. It then goes on to explore the use of street art by a range of activist collectives, including the *Círculo 70*, *Mujeres Creando* and *Insurgencia Comunitaria*. Across these cases, the chapter illustrates the power and importance of political street art in articulating alternatives and bringing marginalised actors and power asymmetries into view.

As Chapter 5 shows, in its early stages Argentine street art was shaped by the import of methods and styles from abroad. However, over the course of the twentieth century, Argentine artists and activists have innovated, appropriated and honed an expansive range of techniques to galvanise and express *la opina de la calle.* This includes graffiti, stencils, murals, mobile silkscreen, street-based performance as well as more hybrid forms of political street art such as *escrache.* Chapter 5 discusses the varying ways that street art can operate as a mode of resistance, exploring three large-scale outpourings of street art which have taken place in Argentina's urban zones: the *Tucumán Arde* intervention of 1968; the *Siluetazo* of 1983 and the outpouring of stencil articulations that accompanied the 2001 financial crisis. It posits that while street art may come to serve a number of important functions during moments of crisis, it too may be facing a crisis of a rather different kind.

Finally, Chapter 6 draws together some of the key themes and arguments advanced in the book. It offers some reflections on street art as a 'contentious performance' that moves between the rational and the non-rational in ways that challenge, contest and recast political relations. It also discusses the ongoing relevance and place of political street art production and reception in a world increasingly saturated with digital images. All in all, the book aims to offer theoretical and empirical advances of use to students and scholars across multiple disciplines including social movement studies, Latin American studies and global politics.

Notes

1 Launched in June 2014, this is probably the largest street art database in the world. As of March 2015 it featured roughly 260 virtual street art exhibits from 34 countries. It gives users the option to browse artworks or hear guided tours.
2 Following Margaret Macintyre Latta (2002), I use the term 'aesthetic play' to refer to attunement to the creating process grounded in the act of making, as explored variously by Bakhtin, Schiller, Dewey and Gadamer.

References

Austin, J. (2010) More to see than a canvas in a white cube: For an art in the streets. *City*. 14 (1–2), pp.33–47.

Batliwala, S. (2008) *Changing their World: Concepts and practices of women's movements*. Toronto: Association for Women's Rights in Development.

Belfiore, E. and Bennett, O. (2008) *The Social Impact of the Arts: An intellectual history*. London: Palgrave Macmillan.

Benjamin, T. (2010) *La Revolucion: Mexico's Great Revolution as memory, myth, and history*. Houston: University of Texas Press.

Bickford, L. (1999) The Archival Imperative: Human Rights and Historical Memory in Latin America's Southern Cone. *Human Rights Quarterly*. 21(4), pp.1097–1122.

Bickford, L. (2000) Human Rights Archives and Research on Historical Memory: Argentina, Chile and Uruguay. *Latin American Research Review*. 35(2), pp.160–182.

Café, R. (2014) London 2012: Banksy and street artists' Olympic graffiti. *BBC News Online*. Retrieved from: www.bbc.co.uk/news/uk-england-london-18946654

Chaffee, L.G. (1993) *Political Protest and Street Art: Popular tools for democratization in Hispanic countries*. Westport, Connecticut: Greenwood Publishing Group.

Goodwin, J. and Jasper, J. (2014) *The Social Movements Reader: Cases and Concepts*. 3rd edn. Chichester: Blackwell Publishers.

Hall, S. (1993) Encoding, Decoding, *in* During, S. (ed.) *The Cultural Studies Reader*. London: Routledge, pp.507–517.

Highmore, B. (2010) Bitter After Taste: Affect, Food and Social Aesthetics, *in* Gregg, M. and Seigworth, G. (eds) *The Affect Theory Reader*. Durham: Duke University Press, pp.118–137.

Johnston, H. (2011) *States and Social Movements*. London: Polity Press.

Johnston, H. and Almeida, P. (2006) *Latin American Social Movements: Globalization, democratization, and transnational networks*. Maryland: Rowman and Littlefield.

Latta, M.M. (2002) Seeking fragility's presence: The power of aesthetic play in teaching and learning. *Philosophy of Education Archive*, pp.225–233.

Mattelart, A. (2008) *Communications/excommunications: an interview with Armand Mattelart, Conducted by Costas M Constantinou*. Retrieved from: www.fifth-estate-online.co.uk/comment/Mattelart-intervie%5B1%5D.pdf

Mendoza, R.G. and Torres, C.C. (1994) Hispanic Traditional Technology and material culture in the United States, *in* Kanellos, N., Weaver, T. and Fabregat, C.E. (eds) *Handbook of Hispanic Cultures in the United States: Anthropology*. Houston: Arte Publico Press, pp.59–84.

Miller, N. (2006) The historiography of nationalism and national identity in Latin America. *Nations and Nationalism*. 12(2), pp.201–221.

Mitchell W.J.T. (1996) What Do Pictures 'Really' Want? *October*, 77. pp. 71–82.

Mutiny (2015) *'Femme Fierce' – My Reflections.* Crick Centre Blog www.crickcentre.org/ blog/femme-fierce-my-reflections/

NPR (2013) *Art Revolution Blooms After Arab Spring.* Retrieved from: www.npr. org/2013/11/07/243720260/arab-spring-artists-paint-the-town-rebel

Oliver, P.E. Cadena-Roa, J. and Strawn, K.D. (2003) Emerging trends in the study of protest and social movements. *Research in Political Sociology.* 12(1), pp.213–244.

Petras, J. and Veltmeyer, H. (2011) *Social movements in Latin America: Neoliberalism and popular resistance.* London: Palgrave Macmillan.

Plato (360 BCE/1976). *The Republic.* (Desmond Lee, Trans.). Middlesex: Penguin Classics.

Project on Ethnicity and Race in Latin America. (2016) *Untitled.* Retrieved from: https:// perla.princeton.edu

Ramos, H. and Rodgers, K. eds (2015). *Protest and Politics: The Promise of Social Movement Societies.* Vancouver: UBC Press.

Rancière, J (2010) 'The Paradoxes of Political Art', in Corcoran, S. ed., (2010) *Dissensus: On Politics and Aesthetics.* London: Continuum.

Rancière, J. (2011) The thinking of dissensus: politics and aesthetics. *Reading Rancière,* London and New York: Continuum, pp.1–17.

Riggle, N. (2010) Street Art: The Transfiguration of the Commonplaces. *The Journal of Aesthetics and Art Criticism.* 68(3), pp.243–257.

Ryan, H. (2014) Indignação! Brazilian street art in its historical context. *The Conversation AU.* Retrieved from: https://theconversation.com/indignacao-brazilian-street-art-in-its-historical-context-27926

Serbin, K. (2006) Review Article: Memory and Method in the Emerging Historiography of Latin America's Authoritarian Era. *Latin American Politics & Society.* 48(3), pp.185–198.

Shaw, R. (2001) Processes, Acts, and Experiences: Three Stances on the Problem of Intentionality. *Ecological Psychology.* 13(4), pp.275–314.

Snow, D. and Benford, R. (1988) Ideology, Frame Resonance and Participant Mobilization. *International Social Movement Research.* 1, pp.197–217.

Snow, D., Rochford, E. Jr., Worden, S., and Benford, R. (1986) Frame Alignment Processes, Micromobilization and Movement Participation. *American Sociological Review.* 51(4), pp.546–481.

Sommer, R. (1975) *Street Art.* Chicago: Links.

Taylor, V. and Van Dyke, N. (2004) 'Get up, stand up': Tactical repertoires of social movements. *The Blackwell companion to social movements*, pp.262–293.

Tilly, C. (2008) *Contentious Performances.* Cambridge: Cambridge University Press.

Tripp, C. (2012) A Lecture on Art and the Arab Uprisings. *Arab Awakening, Open Democracy.* Retrieved from: www.opendemocracy.net/arab-awakening/professor-charles-tripp-on-art-and-arab-uprisings-event-recording

Tripp, C. (2013) *The Power and the People: Paths of Resistance in the Middle East.* Cambridge: Cambridge University Press.

Tschabrun, S. (2003) Off the Wall and into a Drawer: Managing a Research Collection of Political Posters. *The American Archivist.* 66, pp.303–324.

von Clausewitz, C. (1918) *On War*, trans. Col. J.J. Graham. London: Kegan Paul, Trench, Trubner & C.

Weber, J. (1972) Murals as People's Art. *Liberation*, pp.43–49.

Yashar, D. (2015) Does Race Matter in Latin America? *Foreign Affairs.* Retrieved from: www.foreignaffairs.com/articles/south-america/2015-02-16/does-race-matter-latin-america

2 From 'excommunication' to political expression

Conceptualising political street art in Latin America

Traditionally, the term 'excommunication' has been used to describe the suspension or limitation to membership of a religious community or organisation, an act normally undertaken as a mode of censure. However, Armand Mattelart's reworking and extension of the term directs us instead towards the range of factors that can separate persons from the processes, channels and circuits of exchange within society (Mattelart 1996, 2008). Where Mattelart's focus is on the birth and growth of exclusionary technologies, the term is used in an even broader sense here to illuminate the confluence of present and past socio-political processes and events that have combined to distance certain actors and groups from meaningful political dialogue in Latin America. 'Excommunications' in Latin America have resulted from oft-interrelated factors including: poverty, stark and persistent income inequalities; the marginalisation of particular political, social and ethnic groups as legacies of the colonial encounter; as well as authoritarianism and the violent suppression of dissent.

This chapter first explores some of the ways that street art can – and has – enabled agency in the face of 'excommunication'. It then goes on to discuss political street art in the context of the existing body of literature on social movements. It develops a framework for exploring street art as a 'contentious performance' that both draws from and builds upon social movement scholarship in novel ways.

Excommunications: poverty and inequality in Latin America

Throughout the modern history of Latin America, poverty and inequality have emerged as major inhibitors to social, economic and political progress. Today, it is estimated that between 80–100 million people in the region as a whole live in extreme poverty, with approximately half of these located in Brazil and Mexico (World Bank 2013). The largest single group of people in the region is classified as 'vulnerable', existing on $4–10 day and at risk of slipping back into poverty if their fortunes change.

In 2003, *The Economist* described pervasive inequality as Latin America's 'stubborn curse'. In 2012, Christian Aid referred to it as a 'scandal' (Christian Aid 2012). Although between 70 and 90 million people were lifted out of poverty in

Latin America between 2004 and 2014 (World Economic Forum 2016), the region still contains some of the most unequal societies in the world. Based on the Gini index, which is the most commonly used measure of income inequality, 11 Latin American countries featured among the 15 most unequal states across the globe in 2012 (World Bank 2014). Colombia ranked as the most unequal with a score of 53.5, where a score of 0 represents perfect equality and an index of 100 implies perfect inequality. It was closely followed by Brazil (52.7), Panama (51.9), Costa Rica (48.6), Mexico (48.1) and Bolivia (46.6) (World Bank 2014). As of 2014, there were 114 billionaires in Latin America worth a total of $440 billion (Alexander 2014).

The political implications of income inequality in Latin America have been various and complex. On one hand, as Kaufman (2009) highlights, most of the democratic transitions in the region occurred during the 1980s and early 1990s, a period of increasing inequality, and most democracies persisted despite continuing high levels of inequality throughout the 1990s and early 2000s. However, even though a high level of income inequality does not tend to preclude the possibility of formal democracy emerging, it certainly does have a strong bearing on the quality of democracy. At a most basic level, income inequalities have implications for equality of access to education and levels of literacy. Reading, writing and listening skills are 'fundamental to informed decision-making, to active and passive participation in local, national, and global social life, and to the development and establishment of a sense of personal competence and autonomy' (Stromquist 2005). Additionally, as Reuschemeyer (2004) suggests, in highly unequal societies, those in the upper income brackets may be better able to convert their assets into political power and to alter political agendas in their favour. Historically in Latin America, this has perhaps been most apparent in the levying of influence by large-scale landowners. More recently, a spate of corruption scandals across the region point to the undue influence of a new generation of capitalists – magnates, bankers and billionaires – over national political and economic agendas. In a recent report, Christian Aid (2012) comments that despite some progress, 'The economic and political elites and agrarian aristocracy maintain an iron grip on nations' wealth, natural resources, political spaces and media' across the continent.

A great deal of research has also focused on the ways that politicians and policy-makers respond differentially to the views of upper and lower income groups. In some cases, the poor and indigent may be less likely to vote or engage with the formal political process, due to lack of understanding, mobility or even anomie. Yet, income inequalities have also perpetuated damaging forms of *clientelism*, whereby politicians capitalise on income differentials, offering select benefits to the socio-economically deprived in exchange for unquestioning political support.

It is often said that political equality, or the extent to which citizens have an equal influence over the government and its policies, forms the base of democracy. According to Verba (2001), 'One of the bedrock principles in a democracy is the equal consideration of the preferences and interests of all citizens. This is expressed in such principles as one-person/one-vote, equality before the law, and equal rights of free speech.' He continues:

> Equal consideration of the preferences and needs of all citizens is fostered by equal political activity among citizens; not only equal voting turnout across significant categories of citizens but equality in other forms of activity. These activities include work in a political campaign, campaign contributions, activity within one's local community, direct contact with officials, and protest. Equal activity is crucial for equal consideration since political activity is the means by which citizens inform governing elites of their needs and preferences and induce them to be responsive. Citizen participation is, thus, at the heart of political equality (ibid.).

Citizen participation in politics serves a pedagogic and communal function as it brings people, cultures and claim-making practices together. It also fosters legitimacy and confidence in the system, since political decisions are weighed against the sentiments and demands of an active *demos*. Conversely, the absence of widespread citizen participation can be seen to present a direct threat to democratic politics.

The causes of persistent and stark inequality in Latin America have been the subject of much debate. Economist Jeffery Williamson (2009) suggests that the 'stubborn curse' is a twentieth-century phenomenon, brought on during the period of accelerated international commerce and concentration of export revenues coinciding with the *belle époque*. Meanwhile, others have led the charge against the market liberalisations of the 1980s, arguing that the removal of subsidies, privatisation of industries and deregulation of markets as part of the Washington Consensus presented new rent opportunities for the wealthy, while exposing the poor to a range of new insecurities.

Although market liberalisations almost certainly contributed to the growth of the informal sector in the 1990s, it is also crucial to acknowledge longer-term, structural factors at play. Alicia Bárcena, Executive Secretary for the Economic Commission for Latin America and the Caribbean (ECLAC), cited by Christian Aid (2012), suggests that, in fact, 'Inequality has permeated five centuries of racial, ethnic and gender-based discrimination in the region, in societies where people are divided into first- and second-class citizens. It has permeated a modernisation process built on the back of the worst income distribution in the world'. Arguably, the roots of inequality in Latin America lie in the character and process of European colonisation of the Americas, which set a pattern of exploitation and hierarchy amongst indigenous Indians, enslaved and indentured workers from Africa, Asia and Europe.

Across the Americas, the arrival of the Spanish, Portuguese, Dutch, French and British colonists led to the widespread enslavement of indigenous peoples, many of whom perished from hardship or disease. In countries including Brazil, Venezuela, Cuba and the Dominican Republic where the indigenous population had dwindled or been deemed otherwise insufficient for the scale of intended production, colonial officials began to enslave and import Africans to work on the land. This practice was continued by the post-colonial governments in some countries up until the 1880s, when slavery was abolished in the Western Hemisphere. Even after slavery was outlawed, new politico-legal mechanisms were often put in place with the aim of keeping former slaves and new indentured workers on the *fazendas*, *haciendas* and plantations. The Land Law of Brazil (1850) for example, banned the acquisition of land by any means other than cash purchase, in effect reserving the possibility of land ownership for wealthy European migrants and their descendants (DeWitt 2002). Meanwhile, in Peru, Bolivia, Ecuador, Guatemala and other countries where a large indigenous population remained in situ, a variety of galling methods were put into practice to pacify and subjugate, including the expropriation of land, decimation of Indian leadership and measures to trap indigenous workers into debt arrangements which deprived them of their autonomy. In Chile's 'pacification' process of the 1860s the Mapuche people lost large tracts of their land in a series of military campaigns displacing them to the far south of the country (Miller 2006).

Some economic historians have argued for a distinction between Caribbean, Andean and Southern Cone economies in Latin America, pointing out that in Southern Cone countries such as Uruguay and Argentina, the labour diversification strategies pursued during the export boom period of the 1850s resulted in the evolution of comparably more equal societies. Unfortunately, they tend to overlook Argentina's War of the Desert (1879–80), when most of the country's remaining indigenous people were killed to open up the pampas for agriculture (Miller 2006). Nevertheless, post-independence these countries saw a mass inflow of skilled migrants from consolidating liberal democracies in Europe whose presence led to the early establishment of familiar mechanisms for the expression of political voice as well as a high level of political literacy. Many of these voluntary migrants to the Southern Cone lobbied their new governments for investment in education, health and for new legislation to protect workers.

As Ronaldo Munck (2008: 24) puts it, 'Latin America, more than elsewhere perhaps, is a living present which is very much shaped by its history'. Yet, historical experiences differ from country to country too. The region's income inequalities have tended to reflect deeper and more entrenched patterns of hierarchy and division linked to the colonial past wherein certain social groups and communities have been denied equal access to education, housing, land, food, healthcare and political power. Such divisions have also been reinforced over time by the perceptions of difference that emerge between 'self' and 'other'. As Anderson (1983) and others point out, communities are forged (or imagined) by fashioning shared symbols, foundational myths and histories. However, the stories and experiences that unite one group of people also work to distinguish them from the next. Whilst perceptions of difference tend to have 'backward linkages' to position in the class structure, colonial social structures and other intersecting relationships of hierarchy, they also have a certain forward dynamism, motivating particular forms of political thought, action or indeed inaction.

This is an important consideration, for where certain communities and groups have found themselves shut out of formal political processes through the processes outlined above, street art can – and has – emerged as an instrument for claim-making, communicating and expressing views about the social and political context. Chaffee (1993) underlines several relevant functions of street art for impoverished or otherwise marginalised social and ethnic groups in Latin America. He highlights an *identity explanation* for the production of street art, whereby street art acts a vehicle for cultural groups to gain greater visibility and recognition. Here he offers the example of Mexico, where in the 1920s muralists adopted the figure of the Mayan Indian to counter dominant Spanish culture in visions for a New Mexico. He also offers an *alternative media explanation*, suggesting that groups lacking access and representation in mainstream media may be motivated to seek out alternative modes of communication or even to develop their own media systems. Observing that outdoor activities – from shopping, eating, smoking and mingling to listening to music – tend to play a prominent role in the everyday lives of Latin Americans, Chaffee also proposes a *street culture/receiver explanation* for street art production: isn't it logical to establish these alternative media systems in the streets, where people spend so much of their time?

Excommunications: authoritarian excess in Latin America

'Excommunication' in Latin America has not been solely a consequence of structural, economic and ethnic inequalities however. It has also been a product of fear. Since Latin American states gained independence from their Spanish and Portuguese rulers in the late eighteenth and early nineteenth centuries, there have been numerous descents into authoritarianism. From the mid-twentieth century, the Cold War context and the opposing ideological aspirations of the US and Soviet Union strongly influenced the direction of events in Latin America. The rise of a new generation of left-wing revolutionary leaders backed by the Soviet Union from the 1950s led to increased polarisation across Latin America as the United States rallied behind established right-wing and conservative social forces to curb the perceived communist threat. In the mid-1960s there followed a spate of military coups across the region, with the first occurring in Brazil with the ousting of President João Goulart in 1964. Under the auspices of Operation Condor, a secret transnational intelligence sharing network used to target leftist dissidents, union and peasant leaders, the United States offered material support to authoritarian regimes in line with its desire to ensure social control and political influence in its 'own backyard' (McSherry 2002). Across many Latin American states, the period lasting from the mid-1960s to mid-1980s was marked by the presence of especially brutal military regimes that resorted to the widespread intimidation, torture and murder of political dissidents and their families.

In states where military regimes took hold, opposition parties and candidates were often barred from competing for political office and many members of the political left were actively targeted, harassed tortured, exiled or killed. In response, some took their activities underground, drawing inspiration from the Cuban Revolution of 1959, which had demonstrated that a socialist alternative may be possible following a popular revolt and a guerrilla war. As Angell (1966) observes, Latin America's revolutionary movements were largely unsuccessful, with the one great exception of the *Sandinistas* in Nicaragua who overthrew the Somoza government in 1979. These battles between left and right, communism and capitalism, were not just fought in the jungles and hinterlands, they were also inscribed and fought out on urban surfaces – on the pavements, walls and facades of buildings.

Political street art in the authoritarian period(s) was shaped strongly by the prevalence of fear. Corradi and Fagen (1992) provide an account of the various discursive and disciplinary tools used to invoke and manage a system of 'politically-determined fear'. This included a combination of 'actual physical repression, control of the society, propaganda and the omnipresent power of the state' as well as 'disinformation, the absence of the defined rules for the 'war' and the absence of spaces where people could meet and acknowledge the presence of one another' (ibid.: 23). Together these features instilled two basic types of fear: the 'dark room' and the 'dog that barks':

> The first is fear of the unknown, a sense of insecurity about something bad: we know the threat exists but we do not know its exact nature. In classical sociological terms this qualifies as fear of an anomic situation; although the blow or harm is seen as imminent, we know neither whence it comes nor how hard it will strike. The second type of fear is stimulated by known danger: the subject anticipates the harm he or she will suffer, and fear springs from a remembered experience with whose harmful dimensions the subject is completely familiar.
>
> (Corradi and Fagen 1992: 14)

Yet, a prevailing state of fear does not preclude individuals and groups from engaging in various forms of 'everyday resistance'. Drawing on observations from his ethnographic work with Malay peasants, the social anthropologist James C. Scott (1990) has argued that structurally similar modes of domination tend to give rise to some broadly comparable patterns of resistance. In particular, when faced with a repressive government machinery, the threat of exile, torture or even death, members of the public and the political opposition may well be seen to play along with the regime's demands, processes and principles for fear of retaliation. Yet, offstage they will always find ways to question the status quo. Scott's concept of 'infrapolitics' describes the ways in which a whole host of dissenting behaviours may be disguised or veiled to avoid violent retaliation by the authorities. The term deliberately fuses the words 'infrastructure' and 'politics', underlining the role of these largely hidden and often highly creative acts of resistance in perpetuating politics and allowing social movements to germinate. In the chapters that follow, there are various examples of political street art that fit the profile of Scott's 'infrapolitics': anonymised graffiti; the dissemination of stencils, bulletins and flyers as part of an underground media system; even examples of seemingly abstract interventions or aesthetic play can work in ways that break the complicity of silence under authoritarianism' (Chaffee 1993) and display non-conformity with a set of conventions imposed from above.

Where Scott's work focuses more on the experience of 'fear' and resistance at the local level, among those who are effectively 'excommunicated', Corradi and Fagen (1992) provide some useful insights into 'winners' fear' (in other words, the fear experienced by the incoming or incumbent regime). Winners fear, it is argued, stems from 'the trauma experienced before the victory, from [the winners'] perception of how their victory [has] affected the losers, from their suspicion that the repressive mechanism unleashed might become an uncontrollable Frankenstein, from the sense that victory is ephemeral and that the tables might someday be turned on them and the losers will take their revenge' (ibid.1992: 14). Viewed in this way, political street art can – and does – have a strong psycho-political effect on sitting authoritarian regimes. Chaffee elaborates that street art under authoritarian regimes 'connotes an activist, collective sense. In essence, it becomes a form of psychological warfare against the dominant culture and elite and reveals an emerging subterranean movement. This is threatening because it connotes a prelude to an organised opposition, or the existence of one... The act [of producing street art] symbolises a culture of resistance exists that dictators pretend to ignore' (Chaffee 1993: 30). Finally, control over artistic production by authoritarian regimes influences the form art takes (Szemere 1992; Adams 2002). Indications that an active but invisible opposition has survived may well feed 'winners' fear', provoking anxieties that the tables could soon be turned. With this in mind, it is perhaps unsurprising that the Latin American regimes developed an array of engaged strategies for dealing with street art. These ranged from imprisoning artists to systematically whitewashing walls and appropriating the medium for their own uses.

By the mid- to late-1970s it had become apparent that many of the regimes had fallen short of their stated objectives, having overwhelmingly failed to find solutions to the economic and social crises plaguing their peoples. During the 1980s, with the winding down of the Cold War, the United States also shifted its support away from the military regimes, adopting the line that the repressive modes of authoritarian rule were thwarting the consolidation of legitimate government in the region. Starting in the late 1970s and continuing through until the 1990s, many Latin American states underwent a transition from authoritarian to democratic government. The return to civilian rule occurred in Ecuador (1979), Peru (1980), Bolivia (1982), Argentina (1983), Uruguay (1984), Brazil (1985), Paraguay (1989) and Chile (1990) (Huntington 1991; Wiarda 1990). Often described as part of a 'third wave' of democratisation (following Huntington 1993), in this period, a model of formal electoral democracy arrived in all states but Cuba.

Much of the existing scholarship on democratisation in Latin America places an emphasis on the role of elite pacts and processes. For example, Kaufman (1986) and Przeworski (1986) both suggest that democratisation processes tend to begin when disagreements mount within an authoritarian coalition. Stepan (1988) focuses specifically on schisms within the military. Higley and Gunther (1992) meanwhile have been preoccupied with the question of 'what makes democracy stick'. They argue that in states with long records of political instability and authoritarian rule, democratic consolidation requires the achievement of 'consensual unity' among elites – that is, agreement on the value of democratic institutions and respect for democratic rules-of-the-game. The problem with all of these works is that they tend to overlook the impact of opposition actors in general, including mass mobilisations as well as examples of 'infrapolitics' and 'everyday resistance' that work to erode military dominance from the bottom-up. As Mainwaring (1989) rightly states, in reality '[m]ost transitions involve complex interactions between regime and opposition forces from a relatively early stage'. Political street art is one site and source of opposition that can prompt and inspire sudden or indeed more gradual change towards democracy.

Street art as 'contentious performance': strategic claim-making and beyond

As the case studies contained within this book will demonstrate, street art is not the exclusive preserve of a democratic civil society. However, it has often given unique forms of voice to those groups excommunicated from political processes by racial and socio-economic hierarchies, repression and fear. It has played an important role in fostering a more inclusive and democratic politics, by bringing new actors into the fold, facilitating claim-making upon government and enabling forms of expression that go beyond strategic claim-making. Political activists and social movements in Latin America and beyond have long been aware of the political power of street art. And yet, social movement scholars have thus far failed to give any serious attention to it. The section that follows develops a framework for exploring street art as a 'contentious performance' that both draws and builds upon existing social movement scholarship.

Studies of social movements and collective action have proliferated since the 1960s, initially developing along somewhat different trajectories in Europe and the United States before converging in the 'political process theory' (PPT) of the 1990s. PPT developed from critiques of the earlier prevailing views that protestors and other social movement participants were irrational mobs, overwhelmed by a collective senselessness. Political process scholars contended that collective action did not result from the psychological dispositions of the mad and the bad, but rather represented rational responses to political problems. Over time, theorists including Tilly (2008; 2010), Tarrow (1998), McAdam and Snow (1997), outlined five main elements that can help us understand how social movements come into being and what form their actions tend to take. These are:

1   *Mobilising structures*: The organisational strength and resources needed for mounting a campaign, including both formal and informal networks of individuals and institutions.

2   *Political opportunities:* The particular set of opportunities and constraints present within a political context that can facilitate or limit mobilisation.
3   *Framing processes:* The various ways that movements endeavour to construct an issue or claim to give it broad appeal.
4   *Repertoires of contention:* The finite catalogue of claim-making strategies or 'contentious performances' that a movement can draw upon in order to make itself heard.
5   *Protest cycles:* The cyclical rise and fall of contentious politics over time. During a 'cycle of contention' or 'moment of madness' there is heightened conflict across the social system, the will to mobilise spreads, and there is an increased pace of innovation in the specific forms that contention takes.

While these conceptual advances have provided some extremely useful insights into how, why and when mobilisations occur and gather pace, PPT's strong structuralism, its emphasis on rational action, and its search for a series of invariant causal variables to explain social movement emergence and behaviour, have significantly limited its scope (Caren 2007; Goodwin and Jasper 2004). For example, mobilising structures are described by Tarrow (1998: 123) as the elements that 'bring people together in the field'. While there are innumerable factors that could help to cement bonds between individuals and groups in society, such as ideology, ritual, culture and emotion, Tarrow (1998) focuses more specifically on models of organisation, looking at the relative merits of hierarchical versus more horizontal membership structures. Other works in this area have tended to emphasise 'measurables', such as amount of financial resources and size of pre-existing membership base.

Most work on political opportunity has focused heavily on the structure of regimes and cohesion of elite politics. McAdam (1996) outlines the following four main dimensions of political opportunity: (1) the relative openness or closure of the institutionalised political system; (2) the stability or instability of that broad set of elite alignments that typically undergird a polity; (3) the presence or absence of elite allies; and (4) the state's capacity and propensity for repression. Meanwhile, Tilly (2008: 179) explains that, 'Seen from the bottom, regime characteristics become political opportunity structure. Regime openness, coherence of the elite, stability of political alignments, availability of allies, repression and facilitation, and pace of change in those elements define opportunity and threat for potential claimants.' Unfortunately, these approaches largely overlook belief systems as well as the intersecting experiences of class, race and gender discrimination that affect how situations are perceived, how risks are calculated and how they are felt by different groups. After all, '[a]n opportunity not recognised is no opportunity at all' (Gamson and Meyer 1996, cited by Guigni 2009: 365) and feelings of anxiety, fear or even hope can shape the perceptions and behaviours of an individual or even an entire community, driving periods of heightened contention, or 'moments of madness' (Tarrow 1993). These factors have so far received scant attention from PPT scholars but they have been brought into view by more recent scholarship from James Jasper, Jeff Goodwin and Francesca Polletta, among others.

Repertoires of contention and framing approaches meanwhile, represent two of the ways that PPT scholars have attempted to bring agency back in to the debate. Tilly describes the catalogue of actions that a movement can draw upon in order to make itself heard publicly using the theatrical metaphors of 'repertoires' and 'performances'. He suggests that, '[o]nce we look closely at collective making of claims, we see that particular instances improvise on shared scripts. Presentation of a petition, taking of a hostage, or mounting of a demonstration constitutes a performance linking at least two actors, a claimant and an object of claims' (Tilly 2010: 35). These performances clump together to make repertoires of claim-making routines that alter incrementally over time. By invoking the theatrical metaphor, Tilly hopes to call attention to 'the clustered, learned, yet improvisational character of people's interactions as they make and receive each others claims…' (Tarrow 2012: 126). Yet, Tilly's theatre metaphor seems somehow underdeveloped or restricted when contrasted with the work of others like Goffman and Butler. For example, in *The Presentation of Self in Everyday Life* (1959), Goffman argues that life itself is strikingly similar to theatre in that people take on roles and perform certain actions in order to create new realities:

> It is Goffman's claim that if we understand how a contemporary American actor can convey an impression of an angst-ridden Danish prince during a presentation of *Hamlet*, we can also understand how an insurance agent tries to act like a professional operating with a combination of expert knowledge and goodwill. If we can understand how a small stage can be used to represent all of Rome and Egypt in *Antony and Cleopatra*, we can also understand how the Disney Store creates a sense of adventure and wonder in any local mall. Also, if we can understand the process by which two paid actors convince us that they are madly in love in *Romeo and Juliet*, we can understand how flight attendants manage and use their emotions for commercial gain.
>
> (Kivisto and Pittman 2007: 272)

By extension then, political and cultural interventions – signing petitions, demonstrating on the street, and indeed producing street art – are not just learned activities and improvisations qua Tilly, they are 'performative acts' in the sense that Judith Butler explains: they don't just describe the political system, they constitute or enact it. In so far as these enactments alter or challenge the status quo, they give birth to political change. All in all, although Tilly (2008: 21) claims that it is necessary to 'look inside contentious performances and discern their dynamics', his work does not exactly do so. It remains at the meso-level, where repertoire and regime interact and evolve together over time.

In PPT, the concept of 'framing' has been borrowed from Goffman (1974) to emphasise the ways in which movement activists construct their actions and identities to draw support. Although Goffman's original discussion of 'impression management' suggests that '[s]ometimes the traditions of an inidvidual's role will lead him to give a well designed impression of a particular kind and yet he may be neither consciously nor unconsciously disposed to create such an impression'

(Goffman 1959: 6), PPT makes framing into a strategic-rational pursuit. As Benford and Snow (2000: 618) summarise, framing 'denotes an active, processual phenomenon that implies agency and contention at the level of reality construction. It is active in the sense that something is being done, and processual in the sense of a dynamic, evolving process. It entails agency in the sense that what is evolving is the work of social movement organisations or movement activists. And it is contentious in the sense that it involves the generation of interpretive frames that not only differ from existing ones but that may also challenge them.'

The success or resonance of a collective action frame is said to rest on its credibility and salience: are the claims legitimate and do they have a basis in real-world experience? How well do they match up with the priorities, values and ambitions of their target audiences? In practice, as Caren (2007) highlights, framing approaches have been forced to carry most, if not all, of the non-structural drivers of mobilisation, including the mediating influences of culture, ideology and psychology. Even though social movements are generally recognised as cultural innovators within these approaches, culture itself is often treated in a narrow instrumentalist fashion and cultural products are only deemed interesting to the extent that they reinforce a particular collective action frame or ideology. Meanwhile, knowledge and experience are often reduced to disembodied forms of cognition and rationalisation. Viewed through such a lens, political street art begins to look extremely boring, wholly propagandist and functionally indifferent from other 'contentious performances'. Hence, if we really want to 'look inside' political street art and 'discern its dynamics', it is necessary to move beyond the limiting parameters of PPT and adopt understanding grounded in practical aesthetics.

An aesthetic turn for social movement studies?

Even though discussion about the role of culture and emotions in social movements has opened up in recent years, it has not adequately addressed questions of an aesthetic nature. The word 'aesthetics' comes from the Greek word *aesthesis*, which refers to sense perception. Traditionally, the field of aesthetics has been concerned with questions of beauty and taste in art, with aesthetic judgement treated as an autonomous activity that can be objective and separated from everyday life. However, the field of aesthetic enquiry has expanded in recent years with the publication of significant new works about the senses and the re-examination of numerous issues including: the number and individuation of human senses; the mechanisms that underlie sensation in the brain and body; and the links between perception, cognition, and sensory imagining. Where some scholars still use the term aesthetics to denote the perceived formal qualities of 'a thing' or place, others refer to the aesthetic as a distinct field or type of knowledge through which power and resistance can operate. This 'heteronomous' or 'practical' understanding of aesthetics underlines the interconnection between art and the social sphere. In particular, it offers a means of apprehending and intervening in the world via sense-based and affective processes (Bennett 2012).

Werbner, Webb and Spellman-Poots (2014: 1) claim that, 'at their heart, all political ideologies, systems and constitutions are aesthetic systems' founded on a particular distribution of the sensible. For Ranciere, such distributions or orders are composed of *a priori* laws, which condition what is possible to see and hear, to say and think, to do and to make. A given distribution of the sensible therefore yields a kind of disciplinary force and may be upheld in a variety of ways. The physical settings that human beings occupy as part of the 'everyday', for example, give them cues to act in particular ways, conferring meanings and providing some kind of guide through a milieu of information and possibilities circulating in society. In this sense, the settings that political actors occupy form a core part of any 'political opportunity structure', setting the parameters for what is sayable, thinkable or doable in a particular political moment. To demonstrate this, one just has to think about example of the US President in the Oval office. Edelman (1996) underlines how the President meeting his aides in this location 'reminds him and them and the public to whom the meeting is reported, of [the President's] status and authority, just as it exalts the status of his aides and defines the general public as nonparticipants who never get to enter the office'. It follows then, that political settings are never neutral. Rather, they are 'usually staged, contrived and even artificial...[and] this allows them to function as extraordinary, dramatic spectacles which are constructed as intrinsically important' (Chadwick 2001). From a young age, publics are encouraged to believe in the legitimacy of 'legislative halls, courtrooms, executive mansions, and even administrative offices as symbols of government by the people and equality before the law' (Edelman 1996: 77). Hence, city planners past and present have reflected on architectural styles, zoning and more besides, prompted by political concerns about the ways in which the urban topography will impact and feed into the social order.

In Western Europe, it is not unusual to see replication of the Classical Greek and Roman architectural styles in the design of public buildings. Adherence to the principles of harmony, balance, symmetry and monumentality – common to classical architecture – have been understood to contribute to a collective sense of constancy, order and impenetrability that harnesses continuity with powerful empires and polities of the past. In a similar vein, as a part of his excavation of high modernism in *Seeing Like a State*, James C. Scott illuminates how authorities and planners in the twentieth sought to reorder physical environments to reflect ideals of scientific rationality and human mastery over nature. Expending with local and practical knowledge – what the ancient Greeks called *metis* – the transformation of Brasilia and villagisation in Tanzania exemplify an excessive confidence in the principles of 'scientific management' for imposing order on the *demos*. Scott goes on to underline passages in Le Corbusier's proposals for The *Ville Radieuse* which suggest that the dark, crowded and impenetrable slums of Paris provide the ideal conditions for revolutionary organising and thus pose a threat to political stability (Scott 1998). The answer for Le Corbusier was to re-design a capital city without nooks and crannies; an urban centre that provides no place to hide, where alternative political ideas do not have spaces to develop outside of the disciplinary gaze of the state.

In both Edelman and Scott's work there is a strong cognisance that architectural and spatial forms can be instrumentalised for more or less democratic ends. They can be used to help regulate and monitor the populace; they can also be harnessed to stifle dissent and popular mobilisation. In fact, we begin to see how architectural objects and settings do far more than represent an idea or ideology; they can themselves enact politics. Monumental edifices afford extra protection to the state officials that work within them, while urban grid systems generate predictable patterns of movement among the population. Equally though, artists and activists can call attention to these structures and topographies of power through their own appropriation of spaces and materials. Street art, for example, can disrupt the authoritarian aesthetic, creating an 'anti-environment', a transformation of existing visual patterns and cues, through which new ideas, actors and actions can be made visible (see McLuhan 1966).

For many of the adherents to practical aesthetics, 'part of the meaning of artistic talent is the ability to sense feelings, ideas and beliefs that are widespread in society in some latent form, perhaps as deep structures or perhaps as unconscious feelings and to objectify them in a compelling way' (Edelman 1996: 52). Notably, thinkers from Aristotle to Baumgarten, Schiller and Brecht have addressed the full register of human intelligence from reason and logos to sensuality, but the latter has been lost on PPT, with consequences for how social movement scholars have moved to address art-activism. Jasper (1998), Flam and King (2005) and Gould (2009; 2010) are among those that have led calls for greater inquiry into the role of feeling in social movement scholarship. But, as Gould explains, although increasing numbers of scholars accept the idea of 'emotion' as a motivational political force, they tend to elevate the role of cognition, logic and rationalisation in emotional responses. To take one salient example, when discussing the cultural outputs of social movements, Eyermann and Jamison (1998: 21) rely on an explanatory formula that repeatedly and somewhat narrowly 'calls attention to the creative role of consciousness and cognition in all human action, individual and collective'.

For Gould, this preoccupation with rationality is a hangover from the 1970s, when social movement scholars rallied to depathologise protesters, countering dominant depictions of angry mobs and irrational crowds with those of thinking, rational actors. However, the issue runs deeper than that. As Roland Bleiker (2009: 2) explains: 'there is a deep scepticism towards an aesthetic engagement with politics.' This scepticism goes all the way back to the work of Plato, and has shaped the development of the Western philosophic tradition. In his *Allegory of the Cave*, Plato reinforced the view that what was truly real (the world of forms) could only be grasped or known by an enlightened and reasoned mind. The earth and all that was in it was but a poor imitation of the real. Therefore, art, craft – *things perceived with the senses* – could not generate real knowledge. Indeed, they posed a distinct danger to society, by distracting man from truth. In making this distinction, 'Plato separated the body from the spirit, the senses from the mind, and embodied experience from abstract thought. However, [the problem] isn't just that Plato separated these realms in his philosophy but that his separation was not neutral; he granted worth to some ways of knowing and devalued others' (Gonález-Andrieu 2012). Indeed, the devaluation of sensate knowledge and emotion is evidenced throughout many of the most prominent philosophic interventions. As Neta Crawford (2000: 126) relays, Plato and (later) Kant rejected a role for passion in reason; Aristotle could be thought of as an early cognitivist, with his argument that fear works as a certain expectation of undergoing a destructive experience; and, Descartes viewed emotions as 'both biological and cognitive, [but] with emotion following perception'.

But this understanding has not gone uncontested. David Hume rallied against the primacy of reason, arguing: 'Reason is, and ought only to be the slave of the passions, and can never pretend to any other office than to serve and obey them' (Hume cited by Crawford, ibid.). Moreover, the problem as Baumgarten has articulated, is that in privileging abstract models of social life based on disembodied cognition, logic and rational action, we can miss just as much as we gain:

> Indeed, I believe that philosophers can now see with the utmost clarity that whatever formal perfection inheres in cognition and logical truth can be attained only with a great loss of much material perfection. For what is this abstraction but loss? By the same token, you cannot bring a marble sphere out of an irregular piece of marble without losing at least as much material as the higher value of roundness demands. (Baumgarten cited by Askin et al. 2014: 13)

Baumgarten's observation is especially pertinent in the case of contentious politics and political street art, where strategy and cognition are often bound up with and complicated by 'affect'. Gould (2009; 2010), drawing on the work of Brian Massumi describes affect as *nonrationalised* sensation or the corporeal quality of feeling. Affect is deemed distinct from 'emotion', which for Gould describes sensations that have been rationalised, checked against previous sensual experiences and categorised accordingly. Although over time we develop linguistic categories to represent and make sense of our sensory reactions – fear, love, anger, hate – affect is extra-linguistic: it is never fully captured and interned by these words and categories. This makes it a significant site and source of possibility, since the recurring mismatch of existing frames of reference to our felt intensities can drive our quest for new modes of expression and understanding.

For example, in his description of 'moral shock', Bleiker (2009) suggests that at certain historical junctures, moments of crisis and transition, communities, or indeed entire societies may experience a gap or pause in comprehension brought on by the lack of adequate categories for describing and processing the phenomenon at hand. In these instances, non-verbal responses such as painting can enable us to work through our feelings or express compulsions that we cannot yet put words to. Here, the concept of affect makes it possible to consider the ways in which activists are sometimes moved to produce political street art, not in response to some clear strategic imperative, but rather as a non-rationalised form of expression. This ability to capture, process and transfer sensation, mood or feeling in ways that go beyond the categories imposed by prevailing linguistic, cultural and societal codes is what makes aesthetic knowledge unique. As Bennett (2012: 5) argues, affect is the natural medium of aesthetics and it helps us to understand more fully 'what art and imagery *does* – what it *becomes* – in its very particular relationship to events'.

Greater discussion around the role of affect in protest is also vital because it effectively 'retools our thinking about power' (Gould 2009: 27). The affective dimension is an additional field in which and through which power operates. This has implications for the ways we think about both political opportunity and framing processes. On the one hand, it gives us greater space to examine how emotions such as 'anxiety' or 'fear' can emerge and subside, in turn shaping political action and artistic responses. On the other, it also enables us to build a more nuanced picture of how and why some 'schemata of interpretation', action frames, or ideologies are taken up at the expense of others. Gould, for example is focused in particular on the ways in which affect has altered the stakes for AIDS activism. She stresses the ways that affect becomes attached to the discourses of hegemonic power and resistance giving them traction. Similarly, Ty Solomon's illuminating discussion of 9/11 discourses calls for attention to the role of affect. Solomon (2012) argues that most studies examining the social construction or 'framing' of the War on Terror centre on its various narrative strands and seek to show that the understandings which dominate media representations for example, do not constitute an objective condition or truth. However, in the absence of an objective condition or truth, there is little clue as to why certain discourses gain appeal over others. It is here that Solomon outlines the importance of affect in enlivening objects, words and narratives. Quite simply, some *thing, energy* or *state of being* has to be harnessed to make a particular discourse resonate with people. Affect modulation is therefore a form of power, but one that is contingent, hard to observe, challenging to describe and always deployed imperfectly. Affects are volatile in politics, aesthetics and life, but this is why they are the proper – if messy and impure – subject of a practical aesthetics grounded in everyday life and experience (Bennett 2012).

All in all, there are strong grounds to call for an 'aesthetic turn' in social movement theory: to encourage scholars to take seriously the full register of knowledge and experience fostered by aesthetic or sensory engagement in and around contentious politics. Redirecting attention to the role of affect and the sensate field allows us to acknowledge that not all meaningful political and creative activities undertaken by social movements are driven by clear processes of rationalisation and strategic planning. But neither does this imply that they are *irrational* acts borne of collective neurosis. Indeed, the combination of unstructured interviews, observation and archival research that provide the foundation for this book reveal that producing street art enables expressions that 'move back and forth between imagination and reason, thought and sensibility, memory and understanding, without imposing one faculty upon another' (Bleiker 2009: 209). Building on PPT with insights drawn from practical aesthetics makes it possible to go much further in understanding the fluid and complex dynamics of political street art, including not just the strategic, but also the spatial, performative and affective dimensions of the medium which have not received a thoroughgoing analysis in existing scholarly work.

References

Adams, J. (2002) Art in social movements: Shantytown women's protest in Pinochet's Chile. *Sociological Forum.* 17(1), pp.21–56.

Alexander, D. (2014) Meet the Richest Billionaires in Latin America. *Forbes Magazine.* Retrieved from: www.forbes.com/sites/danalexander/2014/03/19/meet-the-richest-billionaires-in-latin-america/#6547a9527738

Anderson, B. (1983) *Imagined Communities.* New York: Verso.

Angell, A. (1966) Party Systems in Latin America. *The Political Quarterly.* 37(3), pp.309–323.

Askin, R., Ennis, P.J., Hägler, A. and Schweighauser, P. eds (2014) *Speculation V: Aesthetics in the 21st Century* (Vol. 5). New York: Punctum Books.

Benford, R.D. and Snow, D.A. (2000) Framing processes and social movements: An overview and assessment. *Annual Review of Sociology.* 26, pp.611–639.

Bennett, J. (2012) *Practical Aesthetics: Events, affect and art after 9/11.* London: IB Tauris.

Bleiker, R. (2009) *Aesthetics and World Politics.* London: Palgrave Macmillan.

Caren, N. (2007) Political Process Theory, *in* Ritzer, G. ed. *Blackwell Encyclopedia of Sociology.* London: Blackwell Publishing. Retrieved from: www.blackwellreference. com/subscriber/tocnode? id=g9781405124331_chunk_g978140512433122_ss1-41

Chadwick, A. (2001) The electronic face of government in the Internet age: Borrowing from Murray Edelman. *Information, Communication & Society.* 4(3), pp.435–457.

Chaffee, L.G. (1993) *Political Protest and Street Art: Popular tools for democratization in Hispanic countries.* Westport, Connecticut: Greenwood Publishing Group.

Christian Aid (2012) The Scandal of Inequality in Latin America and the Caribbean. A Christian Aid Report. Retrieved from: www.christianaid.org.uk/images/scandal-of-inequality-in-latin-america-and-the-caribbean.pdf

Corradi, J.E. and Fagen, P.W. (1992) *Fear at the Edge: State terror and resistance in Latin America.* Berkeley: University of California Press.

Crawford, N.C. (2000) The passion of world politics: Propositions on emotion and emotional relationships. *International Security,* 24(4), pp.116–156.

DeWitt, J. (2002) *Early Globalization and the Economic Development of the United States and Brazil.* Westport, Connecticut: Greenwood Publishing.

Edelman, M. (1996) *From art to politics. How Artistic Creations Shape Political Conceptions.* London: University of Chicago Press.

Eyerman, R. and Jamison, A. (1998) *Music and Social Movements: Mobilizing traditions in the twentieth century.* Cambridge: Cambridge University Press.

Flam, H. and King, D. (2005) *Emotions and Social Movements.* London: Routledge.

Goffman, E. (1959) *The Presentation of Self in Everyday Life.* Garden City, New York: Anchor Press.

Goffman, E. (1974) *Frame Analysis: An essay on the organization of experience.* London: Northeastern University Press.

Gonález-Andrieu, C. (2012) *Taking Back the Aesthetic.* 2012 Princeton Lectures on Youth, Church, and Culture. Retrieved from: www.ptsem.edu/lectures/?action=tei&id=yo uth-2012-01

Goodwin, J. and Jasper, J.M. (2004) *Rethinking Social Movements: Structure, meaning, and emotion.* London: Rowman & Littlefield.

Gould, D.B. (2009) *Moving Politics: Emotion and ACT UP's fight against AIDS.* Chicago: University of Chicago Press.

Gould, D.B. (2010) On Affect and Protest, *in* Cvetkovich, A., Reynolds, A. and Staiger, J. (eds) *Political Emotions.* London: Routledge, pp.18–44.

Guigni, M. (2009) Political Opportunities: From Tilly to Tilly. *Swiss Political Science Review.* 15(2), pp.361–368.

Higley and Gunther (1992) *Elites and Democratic Consolidation in Latin America and Southern Europe.* Cambridge: Cambridge University Press.

Huntington, S.P. (1993) *The Third Wave: Democratization in the late twentieth century* (Vol. 4). Norman, Oklahoma: University of Oklahoma Press.

Jasper, J.M. (1998) The emotions of protest: Affective and reactive emotions in and around social movements. *Sociological Forum.* 13(3), pp. 397–424.

Kaufman, R. (1986) Liberalization and Democratization in South America: Perspectives from the 1970s, *in* O'Donnell et al., (eds) *Transitions from Authoritarian Rule.* Baltimore: Johns Hopkins University Press, pp. 85–107.

Kaufman, R. (2009) The political effects of inequality in Latin America: some inconvenient facts. *Comparative Politics.* 41(3), pp.359–379.

Kivisto, P. and Pittman, D. (2007) Goffman's Dramaturgical Sociology, *in* Kivisto (ed.) *Illuminating Social Life: Classical and Contemporary Theory Revisited,* London: Sage, pp.297–318.

Mainwaring, S. (1989) Transitions to Democracy and Democratic Consolidaition: Theoretical and Comparative Issues. *Kellog Institute Working Papers,* #130. Retrieved from: https://kellogg.nd.edu/publications/workingpapers/WPS/130.pdf

Mattelart, A. (1996) *The Invention of Communication.* Minneapolis: University of Minnesota Press.

Mattelart, A. (2008) *Communications/excommunications: an interview with Armand Mattelart, Conducted by Costas M Constantinou.* Retrieved from: www.fifth-estate-online.co.uk/comment/Mattelart- intervie%5B1%5D.pdf

McAdam, D. (1996) Conceptual origins, current problems, future directions, *in* McAdam, D., McCarthy, J. and Zald, M. (eds) *Comparative Perspectives on Social Movements: Political opportunities, mobilizing structures, and cultural framings.* Cambridge: Cambridge University Press, pp.23–40.

McAdam, D. and Snow, D.A. (1997) *Social Movements: Readings on their emergence, mobilization, and dynamics.* Los Angeles, California: Roxbury.

McLuhan, M. (1964) *Understanding Media: The Extensions of Man.* New York: McGraw Hill.

McLuhan, M. (2005 [1966]) The Emperor's Old Clothes. In G. Kepes (ed.), *The Man-Made Object* (pp. 90–95). New York: George Brazillier Inc. Reprinted in E. McLuhan and W. T. Gordon (eds), *Marshall McLuhan Unbound* (20). Corte Madera (California): Gingko Press.

McSherry, J.P. (2002) Brazil: The Hegemonic Process in Political and Cultural Formation. *Latin American Perspectives.* 29(1), pp. 38–60.

Miller, N. (2006) The historiography of nationalism and national identity in Latin America. *Nations and Nationalism.* 12(2), pp.201–221.

Munck, R. (2008) *Contemporary Latin America.* Palgrave Macmillan.

Przeworski, A. (1985) *Capitalism and Social Democracy.* Cambridge: Cambridge University Press.

Przeworski, A. (1986) Some Problems in the Study of the Transition to Democracy. In O'Donnell et al. (eds) *Transitions from Authoritarian Rule*, Part 3, pp.47–63.

Reuschemeyer, D. (2004) Addressing Inequality. *Journal of Democracy.* 15(4), pp.76–90.

Scott, J.C. (1990) *Domination and the Arts of Resistance: Hidden transcripts.* New Haven: Yale University Press.

Scott, J.C. (1998) *Seeing like a State: How certain schemes to improve the human condition have failed.* New Haven: Yale University Press.

Solomon, T. (2012) 'I wasn't angry, because I couldn't believe it was happening': Affect and discourse in responses to 9/11. *Review of International Studies.* 38(4), pp.907–928.

Stepan, A. (1988) *Rethinking Military Politics: Brazil and the Southern Cone.* Princeton: Princeton University Press.

Stromquist, N. (2005). The political benefits of adult literacy. Background paper for EFA Global Monitoring Report 2006. Retrieved from: unesdoc.unesco.org/images/0014/001461/146187e.pdf

Szemere, A. (1992) The politics of marginality. A rock musical subculture in socialist Hungary in the early 1980s, *in* Garofalo, R. (ed.) *Rockin' the Boat. Mass Music and Mass Movements.* Boston, Massachusetts: South End Press, pp.93–114.

Tarrow, S. (1993) Cycles of collective action: Between moments of madness and the repertoire of contention. *Social Science History.* 17(2), pp.281–307.

Tarrow, S. (1998) *Power in Movement.* 2nd edn. New York: Cambridge University Press.

Tarrow, S. (2012) *Strangers at the Gates: Movements and states in contentious politics.* Cambridge: Cambridge University Press.

Tilly, C. (2008) *Contentious Performances.* Cambridge: Cambridge University Press.

Tilly, C. (2010) *Regimes and Repertoires.* Chicago: University of Chicago Press.

Verba, S. (2001) Thoughts about Political Equality: What Is It? Why Do We Want It?. *Inequality Summer Institute, Harvard University.* Retrieved from: www.russellsage.org/sites/all/files/u4/Verba.pdf

Werbner, P., Webb, M. and Spellman-Poots, K. Eds (2014) *The Political Aesthetics of Global Protest: The Arab Spring and Beyond.* Edinburgh: Edinburgh University Press.

Wiarda, H.J. (1990) *The Democratic Revolution in Latin America: History, politics, and US Policy.* New York: Holmes & Meier Publishers.

Williamson, J (2009) Latin American Inequality since 1941. *Vox.* Retrieved from: www.voxeu.org/article/latin-american-inequality-1491

World Bank (2013) *Inequality in Latin America falls, but challenges to achieve shared prosperity remain.* Retrieved from: www.worldbank.org/en/news/feature/2013/06/14/latin-america-inequality-shared-prosperity

World Bank (2014) *GINI index (World Bank estimate).* Retrieved from: http://data.worldbank.org/indicator/SI.POV.GINI?order=wbapi_data_value_2012+wbapi_data_value&sort=desc

World Economic Forum (2016) *Inequality Still Presents a Big Challenge for Latin America.* Retrieved from: www.weforum.org/news/inequality-still-presents-big-challenge-latin-america

3 *'Tupinaquim ou Tupinãodá?'*
Rethinking street art in Brazil

Street art decorates every conceivable surface and space within the sprawling urban zones of modern Brazil, providing something of a visual feast for urban residents and visitors. As Manco, Lost Art and Neelon claim in the Preface to their 2005 photo-book *Graffiti Brasil*, the country 'boasts a unique and particularly rich graffiti scene, which in recent years has earned it an international reputation as the place to go for artistic inspiration'. In 2014, the Huffington Post released its *26 Best Cities In The World To See Street Art*, with São Paulo, Brazil ranking in at no. 2. The international popularity and appeal of Brazilian street art has led to a surge in the publication of books, art-zines and even street art tours, many of which offer a cursory examination of street art's history and evolution in Brazil. Many of these sources inaccurately link the birth of Brazilian street art to the import and adoption of hip-hop subculture and associated practices during the 1980s. While it is true that the production of street art, and particularly the distinctive *graphite* style,[1] accelerated at an unprecedented pace from the mid-1980s, this story neglects much of the history and politics surrounding the evolution of Brazilian street art.

Thus far, the only serious and systematic attempt to document the history, evolution and range of political street art in Brazil has been Lyman G. Chaffee's cross-comparative study of *Political Protest and Street Art: Popular Tools for Democratization in Hispanic Countries*. In this text, Chaffee follows the development of street art from the poster campaigns of the interwar period through to the protest graffiti of the 1980s. Chaffee provides a rather general account, noting several challenges to the study of street art in Brazil, including the wide geographical spread of the country; the variation of cultural influences between the North and South of Brazil, as well as great disparities in levels of literacy and access to resources.

This chapter utilises Chaffee's account as a springboard for exploring the history and politics of Brazilian street art through periods of authoritarian and civilian rule. It builds on his work by providing new examples and analyses from poster campaigns and murals to the birth of *pixação*. In particular, the chapter tells the story of *Grupo Tupinãodá*, São Paulo's very first street art collective. *Tupinãodá* formed part of an as yet understudied 'first generation' of *graffiteiros* who emerged in the authoritarian period and used their art to intervene in a context marred by the circumscription of political rights and freedoms, repressive police practices and the inadequate, even arbitrary, application of the rule of law.

Tupinãodá's subversive interventions were the very first to adorn the walls of the *Beco do Batman* and *Paulista/Rebouças* tunnel in São Paulo – painting zones now famous internationally among street artists, tourists and reporters. The group's work and their testimonies offer insights into the authoritarian experience and in particular, the conditions and risks attached to the expression of dissenting views in this period. This chapter underlines the political value of *Tupinãodá's* street art, which offered a sustained counter-discourse and means of re-imagining civic and social life during the dictatorship. Moreover, by appropriating prominent urban spaces and experimenting with large-scale abstract designs their works also set the stylistic and spatial parameters for much Brazilian street art to come.

'Street poster art' and the Paulista War

Although it is claimed that 'Brazil has a graffiti heritage dating back to indigenous rock carvings' (Manco, Lost Art and Neelon 2005: 13), the interwar period serves as the entry point for the study of Brazilian street art in this book. It is here that the first documented and archived examples of political street art can be found. Beginning with the period of corporatist rule under Getulio Dornelles Vargas, street art forms were utilised by incumbent (pro-system) and oppositional (anti-system) political forces. Parties and alliances from across the ideological spectrum used evocative campaign posters to frame their claims, counter-frame their opponents and mobilise resources (financial and otherwise).

Described by Goffman (1974) as '*schemata of interpretation*', collective action frames are used purposefully by political actors in order to focus the attention of a target audience on specific dimensions of an issue, or shape their understanding and beliefs in specific ways. Until quite recently, the analysis of collective action focused mainly on structure and process, with the role of cultural dynamics massively downplayed (Jasper 1998; Zald 1996; Morris 2000). A range of scholars have sought to correct this, by highlighting how shared cultural understandings, common historical reference points and popular symbols may provide a basis/ opportunity for bringing people together beneath one banner. Hence, it is not altogether surprising that certain images and motifs become pervasive in political posters over time. As Seidman (2008: 164) highlights:

> symbols common in the campaigns that political parties have conducted around the world include snakes and octopi, to be combated; lighthouses and the sun, to denote a 'better tomorrow'; the cross to represent a party's religious values or 'martyrdom'; raised fists to demonstrate defiance; as well as a plethora of eagles and other animals, 'V's for victory, and of course, flags.

Yet, the challenge for poster makers is not just to deploy familiar symbols but to use them in ways that have *resonance*. One way of doing this is to successfully orient them towards action within particular contexts; another is to fashion them at the intersection between a target population's culture and the values and goals of the party or movement (Tarrow 1998). These practices can be seen clearly in the examples of street poster art discussed below.

The interwar period in Brazil was marked by internal rivalry between modernisers and the landed powers of the *Republica Velha* (Old Republic). The latter, largely plantation owners and 'coffee barons', had long been engaged in election rigging, machine politics and had managed to perpetuate a *café com leite*[2] pattern of advantage, enabling the two large coffee-producing states of São Paulo and Minas Gerais to alternate the seat of the national presidency between themselves. Coffee exports had been the bedrock of the Brazilian economy for some time but, in the 1920s, the export market 'began to lurch from crisis to crisis' (Chasteen 2006: 233) as overproduction placed downward pressure on prices. Over the same period and having grown increasingly angry with the patrimonial configuration of Brazilian politics, junior army officers or *tenentes* initiated a series of revolts across the country. The Great Depression of 1929 caused coffee prices to plunge further, destabilising the position of the old elite and revealing the vulnerability of the Brazilian economy. In the 1930 presidential elections, Getulio Vargas, then governor of the rising state of Rio Grande do Sul – a non-coffee producing state – stood against the candidate from São Paulo, Julio Prestes. Vargas campaigned with the argument that Brazilian politics should be re-shaped to support national development and industrialisation and he attracted a strong support base, including: *tenentes*, manufacturers, and the *Aliança Liberal* (Liberal Alliance) – a coalition uniting landed classes from the states of Minas Gerais, Rio Grande do Sul and Paraíba. Electoral managers counted ballots in favour of Prestes, but shortly thereafter, Vargas assumed power in a bloodless *coup*.

Attempting to instruct and encourage support for Vargas' rule and his proposed reforms, the *Aliança Liberal* produced a series of sophisticated multi-coloured posters. According to Chaffee (1933: 132), some of the posters contained extensive statements of solidarity or political intent, others attempted to personalise Vargas through images, while others still voiced issues of concern (Chaffee 1993). In one example, an *Aliança Liberal* poster depicted three fatigue-clad figures on horseback, representative of Minas Gerais, Rio Grande do Sul and Paraíba. Each of the figures held his respective state flag. The figure on the far right of the image held the state flag of Paraiba, inscribed in bold with the phrase, 'NÉGO'. Translating as 'I refuse' or 'I deny', the flag recalled the Paraíba's state governor João Pessoa's refusal to support the candidacy of *paulista*, Julio Prestes. Notably, Pessoa was assassinated in 1930 and the black of the flag is said to symbolise the mourning of the *paraibanos*. Behind the three figures is a rising or setting sun, with brilliant red rays extending from the poster's edges and leading the viewer's line of sight towards a distant horizon. In Brazil, as in many other countries and cultures, the symbol of the sun on the horizon is popularly associated with ideas of hope, change and possibility.

Following the coup, political forces from the state of São Paulo also utilised posters to mobilise an opposition and to counter-frame Vargas and the *Aliança*. Chong and Druckman (2013) define a counter-frame as a frame that opposes an earlier effective frame. Competing frames allow individuals to evaluate the

Figure 3.1   Revolutionary Poster, 1932.
Author Unknown. https://commons.wikimedia.org/wiki/File:Cartaz_Revolucionário_1.jpg#/media/
File:Cartaz_Revolucionário_1.jpg

relative strengths of alternative positions and ideologies and compare them to their own priorities and values. In this case, posters appeared in São Paulo projecting slogans such as: '*Em defesa da Constituição*' (In defence of the Constitution), attacking Vargas' autocratic style of government and underlining his disregard for the rule of law. Another read '*Abaixo a dictadura*' (Down with the dictatorship) and depicted in caricature a small figure resembling Vargas in the grips of a giant *bandeirante*. The flag of São Paulo state frames the image. The 'bandeirantes' or '*followers of the flag*', were members of the sixteenth to eighteenth-century South American explorations known as '*bandeiras*'. The original purpose of the bandeiras was to capture and force indigenous people into slavery, but later expeditions began to focus on the acquisition of mineral wealth. *Bandeirantes*, most of whom were from São Paulo, were responsible for expanding Portuguese territory from the Tordesilhas Line to roughly the area occupied by today's Brazil. The poster attempts to belittle Vargas while emphasising the heroism, strength and resilience of São Paulo state, embodied in the muscular figure of the *bandeirante*.

Tensions between the *Aliança* and the *paulistas* escalated in May 1932 when four protesting students from São Paulo were killed by government troops. A movement named MMDC (invoking the initials of the surnames of each of the students: Martins; Miragaia; Dráusio and Camargo) emerged to pressure the provisional government headed by Vargas to abide by a new Constitution. Two months later, social unrest mounted into a full-scale uprising, known variously as the Constitutionalist Revolution or the Paulista War. As fighting ensued, both sides continued to use political posters for instrumental ends. The *paulistas* used them to raise funds for the war effort. One *paulista* poster featured the slogan:

Figure 3.2   MMDC Poster, 1932.
Author Unknown. https://pt.wikipedia.org/wiki/M.M.D.C.#/media/File:Cartão_Postal_do_MMDC.jpg

'*para o bem de São Paulo*' (for the good of São Paulo), accompanied by the image of a woman's hand dropping a gold ring – presumably her wedding band – on to a collection plate. The message here was one that celebrated personal sacrifice and loyalty to the state of São Paulo. The religious undertones are also clear: women are asked to give to the state as freely as they would to the Church.

In these examples, it is possible to see how visual components have been engineered to tap into and resonate with popular and local symbols, histories and cultural practices. Yet, as Tschabrun (2003: 315) argues, in 'the best political posters, the text and graphics work together to express meaning as an intertwined, symbiotic whole'. In each of the cases above it is clear that language is used to pin down meaning, to bring interpretations of the images in line with a particular narrative, discourse or ideology. Hence, '[t]he visual vocabulary of political posters is simultaneously rich and limited'. On the one hand, poster producers are 'constantly recycling, reinterpreting, and transforming a large but restricted body of icons and images' Tschabrun (2003: 315). On the other, the accompanying words and slogans set the parameters for audience engagement and interpretation. In addition to a strong 'symbiosis of word and image', Christine Nelson and Joel Rutstein (1995) suggest that an impactful political poster should also attack the viewer's emotions. That is to say, it should jolt, jar, or rouse the viewer into action. In the absence of testimony or audience response data, however, it is hard to establish the precise impact of posters on viewers themselves.

Yet, they can provide other kinds of insights relevant to social and political history. Firstly, when 'studied for the messages they communicate by the juxtaposition of words and images' (Tschabrun 2003: 305–306), political posters may bring the political, historical and cultural constellations that have generated them into sharper view. The cross-section of posters explored above provides insights into the gendered assumptions that underpinned political participation in the 1920s and 1930s. The characters that feature most visibly on either side of the campaign – the *tenentes*, the *bandeirante* and Vargas – are all men. In São Paulo, the tall, strong and imposing figure of the *bandeirante* is held up as an archetype of masculinity and competent government. Women, meanwhile, are relegated to the status of passive and faceless financiers, urged to sacrifice their most prized personal possessions for the victory of the state in a formula that resonates strongly with the call of feminist scholars: 'the personal is political.'

Secondly, the medium itself can be probed for insights into socio-economic conditions and lifestyles. Seidman (2008: 13) explains that in countries where illiteracy has been high, posters were – and in some cases still remain – an important communicative medium. In Brazil, posters and broadsides sited in public places provided a useful way of conveying information to the *demos*. Often, the slogans would be read aloud by those who were literate. In this way, posters had a dual role in public education: not only did they convey information and choices related to the politics of the day, their proliferation and presence made reading a desirable skill and sparked greater public interest in schooling.

Yet, campaign posters have also been limited in their ability to facilitate a truly democratic and participatory model of politics. In his history of the poster,

Rickards (1971) writes that in less literate societies, the printed word has had an authority of its own, providing a direct point of contact between ruler and ruled. In no uncertain terms it has been an instrument of power. As Tschabrun (2003: 303), summarises: 'Brash and aggressive, raw and yet often poignant, political posters urge, instruct, encourage, and exhort... they share with banners, flyers, signs, and promotional materials of all kinds the purpose of communicating instantly, effectively, and powerfully.' However, the mass production of posters requires specialised equipment, meaning that it was – and is – quite often cost-prohibitive for the poorest and most vulnerable in society to express their interests in this way. Thus, in the period leading up to the Constitutionalist uprising, political posters became the mainstay of political elites and well-financed parties representing the interests of the new urban bourgeoisie, *tenentes* and landowners hostile to the *paulistas*. Poorer members of society – particularly in rural areas – were *spoken to* via political posters, but they had few, if any, opportunities to *talk back* to power. Still, many of the poor and working classes responded positively to Vargas, whose proposals offered 'better working conditions, job security, and opportunities for subsidized housing' (Levine 1998: 10).

The *Paulista* uprising lasted only three months before being supressed by federal forces. With this defeat under his belt, Vargas oversaw the design of a new constitution in 1934 which reorganised aspects of the political system, changing the structure of the legislature and instituting several electoral reforms, including women's suffrage, a secret ballot, and special courts to monitor elections.

For seven years, Vargas ruled as a more or less constitutional president in a country filled with new political energies and actors (Chasteen 2006). However, in 1937 he circumvented constitutional limitations on his re-election, creating a new politico-economic system called the *Estado Nôvo* (New State). The birth of the *Estado Nôvo* ushered in a period of highly authoritarian government, modelled in the image of Antonio Salazar's corporatist state in Portugal. Correspondingly, 'modernization, political centralization [and] industrialization' (Fleischer 2005: 472) were bedrock principles. All political parties were effectively dissolved until 1944, thus limiting opportunities for an opposition to organise. State autonomy was severely eroded with appointed federal officials or *'interventors'* replacing state governors and patronage flowing vertically from the President to those he favoured.

Chasteen (2006: 233) notes that the *Estado Nôvo* 'spawned dozens of new government boards, ministries and agencies, a bit like the alphabet soup agencies of FDR's New Deal, to further the Nation's common goals and welfare. National councils and commissions were created to supervise railroads, mining, immigration, school textbooks, sports and recreation, hydraulic and electrical energy, and so on'. In the context of this national political project, the state exerted new assimilationist pressures: immigrants were told to speak Portuguese, racial mixing was encouraged and Brazilians were told to embrace their African roots. The samba, an afro-Brazilian dance with roots in the dance parties of slaves and former slaves in the rural areas of Rio, became accepted as the country's national dance and was 'vigourously promoted by the mass media of the *Estado Nôvo*' (ibid.). Vargas also appointed key figures of the Brazilian modernist movement

Figure 3.3   Pro-Vargas campaign poster, 1938.

Author Unknown. https://pt.wikipedia.org/wiki/Getúlio_Vargas#/media/File:Propaganda_do_Estado_Novo_(Brasil).jpg

into the state apparatus. Among these were the composer Heitor Villa-Lobos whose folkloric-come-classical compositions drew on Oswald de Andrade's *'Manifesto Antropófago'* (Cannibalist Manifesto) of 1928, which argued that Brazilian artists ought to – metaphorically speaking – 'cannibalise' European art and combine it with distinct native and African influences of Brazil to create a unique national art movement.

A opinião da rua: from handbills to *pichação*

By 1945, mounting societal disillusion with the *Estado Nôvo* led the government to relax its restrictions on political activity. First, censorship of the press was wound down and later came the announcement of congressional and presidential elections. This long-awaited democratic opening generated a great deal of new expressions on the street as various groups sought to project the possibility of political alternatives and communicate their frustrations through publicly accessible social commentary. Competing parties vied for electoral support using posters, handbills, *folhetos*,[3] painted slogans and symbols on the pavements; and, a rich and varied body of myths, legends, anecdotes, and attitudes developed around the figure of Vargas. Members of the urban poor also took to the streets to express their views, giving birth to the tradition of *pichação* in which people would paint with improvised materials such as tar.

Vargas permitted newly formed parties to make preparations for a national election in 1945. However, it was by no means clear that he intended to give up power. A group of staunch Vargas supporters, known as the *queremistas* (named for their slogan *'Queremos Getúlio'* or 'We want Getúlio') embarked on a vocal and highly visible campaign on his behalf. They became prolific producers of *folhetos* and street art posters on Vargas, framing him as *'o pai dos pobres'* (the father of the poor). The *queremistas* – many of whom had been newly enfranchised through Vargas' reforms – were motivated by a natural sentiment of admiration and they developed 'the Sebastianist theme that Vargas would return' (Lauerhass jr. 1979: 80). Great emphasis was placed on Vargas' championing of the poor and underprivileged; 'his patriotism, morality, and good deeds were couched in a style and in symbols that were more acceptable to the rural masses' (ibid.).

As the political playing field opened, the *Partido Comunista Brasileiro* (*Brazilian Communist Party or* PCB) and the labour movement also rallied behind Vargas. The PCB emerged as the fourth largest political party in this period and, according to Chaffee (1993), it became the most prolific producer of street art. Initially, the PCB published and disseminated posters reading *'Constitucionalismo com Getúlio'* (Constitutionalism with Getulio). Rather resourcefully, the PCB also began to use paint to emblazon paving slabs in urban areas with the hammer and sickle, perhaps the most well-known symbol of the international communist movement, representing the unified force of the industrial proletariat and agricultural workers. While the PCB were most active in urban, industrialised areas, the allied *ligas camponesas* (peasant leagues) began to employ graffiti to enhance their visibility in rural and agricultural areas.

Student groups meanwhile rallied against Vargas, utilising a combination of public demonstrations and poster campaigns. In March 1945, a student demonstration in Recife was intercepted by police officers, leading to the death of the president of the Pernambuco Students Union. In response, students from across the country organised a week-long *anistia* (amnesty) campaign, demanding an end to political violence and freedom for Vargas' political prisoners. At around the same time, the *União Democrática Nacional* (National Democratic Union, or UDN) began to mount a campaign on behalf of one of Vargas' opponents, the former *tenente*, General Eduardo Gomes. The UDN presented Gomes as a foil to Vargas, who was viewed by his opponents as inconsistent, demagogic and untrustworthy. The UDN utilised a combination of murals, posters and painted slogans to highlight Vargas' wrongdoings while celebrating Gomes' achievements. For instance, the slogan *'Lembrar 1937'* (Remember 1937), painted on walls and urban edifices, instructed voters to recall the constitutional crisis of 1937, in which Vargas extended his rule. Gomes was one of only two survivors from 'Copacabana Fort', the first revolt undertaken against the politics of the Old State in 1922. The UDN thus sought to frame him as a reliable, consistent and courageous leader by recalling his leadership during the revolts and his respect for the laws of the land. In 1945, the party commissioned a huge mural in Belo Horizonte, depicting Gomes with a book in his hand. The title read, *Constitutions*. UDN posters portrayed Gomes as the 'true' peoples' candidate, reading: *'Votar em Eduardo Gomes'* (Vote for Eduardo Gomes), *'para as pessoas, para a república e pela patria'* (for the people, for the republic and for the fatherland).

As it happened, Vargas and Gomes were both beaten in the polls by Eurico Gaspar Dutra. Dutra was Vargas' former Minister of War and had played an instrumental role in creating the 'state of emergency' in 1937 that allowed for the birth of the *Estado Nôvo*. Having won by a handsome margin, Dutra assumed the presidency in January 1946 and remained in office until 1951. During this time, the Brazilian government had a much closer relationship with the United States. Dutra signed The Inter-American Treaty of Reciprocal Assistance (or Rio Treaty) agreeing that Brazilian forces would deploy in the event of an attack directed towards a state in the Western Hemisphere (see Hilton 1979) and he followed a programme of economic liberalisation that was warmly welcomed by US interests. Dutra's government grew markedly more suspicious of the PCB, which had amassed a large following. In 1946 Dutra enacted a new constitution, which in 1947 was used as a basis for withdrawing the registration of the PCB and expelling all communists from office. Following criticism from Soviet media outlets, Dutra cut diplomatic ties with the USSR entirely. These moves prompted left-wing indignation, which paved the way for Vargas' return to power in 1951. Vargas was reinstated by direct election, forming a coalition cabinet representing all of the major parties. However, inflation, high costs of living, and an increase in underground Communist activity spurred increasing frustrations among the military. Following the accidental death of an air force officer during the attempted assassination of an editor of an anti-Vargas newspaper, Vargas was deposed by the military on 24 August 1954. He committed suicide hours later, becoming one of the most controversial characters in Brazilian history (Fleischer 2005). Variously loved and despised as a dictator, industrialist, socialist, corporatist and populist, Vargas' death cast 'a long shadow' over Brazil (Davila 2006).

As we have seen, in the late 1940s the political playing field had opened up considerably, bringing a raft of new voices and actors to the fore to challenge the status quo. New parties, leagues and alliances – initially devoid of resources – innovated new, low-tech ways of getting their messages out to the populace. Notably, in this period 'the people' also began to speak back in creative ways. Improvising with available resources, impoverished, disillusioned and disenfranchised Brazilians began to turn to forms of ad-hoc wall-writing. As Manco *et al.* (2005: 13) document:

> In the mid 20th century, a growing urban population expressed their opinions through 'wall writings' – political messages written with tar. By the 1940s and '50s, these political writings, known as *pichação*, were commonplace. They were often written in response to the slogans painted by the political parties across the streets.

Distinct from its current form,[4] early *pixação* (or *'pichação'*), which comes from *piche*, the Portuguese word for tar, consisted in the most part of phrases and interjections scrawled hastily on walls, doors or windows in the growing urban centres of São Paulo and Rio de Janeiro. By the 1950s, *pixação*, had become the key medium for the poor and dispossessed, with the city walls providing perhaps the very first accessible platform for marginalised groups to communicate their social and economic concerns to the politicians and broader populace in an unmediated fashion.

For a time, these localised civilian interventions served to neutralise the power of the political propaganda and pluralise *a opinião da rua* (the opinion from the street). However, during the early 1960s, 'pro-military groups also took their campaigns to the city walls' (Chaffee 1993: 133). These groups employed painted slogans, seeking to capitalise on the grass-roots aura of *'piche'* and appropriate it for their own ends. Military campaign activities increased in response to João Belchoir Marques Goulart's appointment as president in 1961. Well known for his moderate left-wing views, Goulart was distrusted by the military and by conservatives who maintained that his sympathies with Fidel Castro's Communist regime made him a threat to national security. Negotiations between Goulart and the military establishment resulted in an agreement whereby a parliamentary system would be instituted to curb Goulart's powers. By 1963 however, Goulart had managed to reinstate a presidential system of government and he issued a series of decrees setting low-rent controls, redistributing land, nationalising oil refineries and limiting the repatriation of profits by large multinational companies.

Pro-military groups became increasingly disgruntled and the names of faction leaders began to appear on the streets, liberally and frequently scrawled in a way that reflected 'an increasingly restless military' (Chaffee 1993: 133). As Chaffee puts it, military groups were 'testing the political waters, jockeying among themselves, advancing one faction and leader over another, and heightening their profile to test public sentiments' (ibid.). This culminated in 1964 when Goulart was overthrown by the military, ushering in over two decades of authoritarian rule.

The 21-year period of authoritarian rule was characterised by varying levels of censorship, human rights abuses and arbitrary implementation of the rule of law. Beginning in 1964, General Humberto Castelo Branco implemented a series of supra-constitutional measures known as *Ato Institutionais* ('Institutional Acts' or AIs), which in effect entrenched military control. The implementation of AI-1 enabled any politician who was perceived to pose a viable threat to the regime to be removed from office. AI-2 abolished the multi-party system, legalising only the government-backed *Aliança Renovadora Nacional* (ARENA) and a new, *Movimento Democrático Brasileiro* (MDB), which served initially as an umbrella organisation for the broad range of liberals, leftists, moderates and conservatives who had been drawn together from outlawed parties and had managed to retain their jobs by distancing themselves from the more critical leftist political currents and characters (Koonings and Krujit 1999). From 1965 onward, the President, governors and some mayors were elected indirectly through the Congress, which in practice was controlled by the military regime.

From the 'years of lead' to *Tupinãodá*

For a short time under Branco's successor, Marshal Artur da Costa e Silva, non-violent protests were tolerated. Marches and demonstrations succeeded in uniting quite diverse sectors of society, all opposed to the military government, its ideology and practices. Politicians from the legal and illegal opposition attempted to consolidate an anti-authoritarian alliance, by the name of the *Frente Ampla* (Broad Front). Realising that there was little possibility of ousting the military through direct or violent forms of confrontation, *Frente Ampla* framed their objectives in terms of *re-thinking* the relationship between the military and civil society. They began producing and disseminating their articles and manifestos via the *Tribuna da Imprensa* newspaper before being outlawed on 5 April 1968. Later in the year, a '*Passeata dos cem mil*' '(March of the 100,000') took place in Rio de Janeiro, spurred by the death of student Edson Luís at the hands of the police. The shock of Luis' death brought together the student movement and elements of the old left as well as the middle class and the Catholic Church, both of which had initially thrown their support behind the military's drive for a more ordered society. The march was led by a single large banner with the hand-painted slogan '*Abaixa a dictadura – Povo no poder*' (Down with the Dictatorship – People in power) and lasted for three hours, without being dispersed by the police. However, feeling increasingly cornered as a result of the year's eruptions and unrest, in December 1968 the regime shut down parliament entirely. Costa implemented AI-5, requiring that all media content be reviewed by government agents prior to publication. AI-5 also removed key political rights, paving the way for political prisoners to be subjected to military law without *habeas corpus* (Koonings and Krujit 1999). In most areas of civil and political governance, Costa assumed unlimited powers.

The implementation of AI-5 was followed by a series of political purges, during which many artists, academics and politicians were arrested, tortured and/or forced into exile. The brutal repression of workers' strikes at Osasco and Contagem in 1968 effectively silenced the labour movement (Mainwaring 1986). Other targets of military brutality included members of the emergent *Tropicalia* movement (or '*Tropicalismo*'), which had been producing politically engaged music and performances since the coup in 1964, and the theatre director Augusto Boal. Inspired by exiled Brazilian educator Paulo Freire, Boal had begun experimenting with popular, integrative theatre in Rio de Janeiro's public spaces during the 1950s. His aim was to open up channels of two-way conversation – or, new democratic spaces – among the populace using applied drama as a tool. Boal developed a theatrical process whereby audience members could halt a performance at any point to suggest different actions for the actor, who would then carry out the audience suggestions. Through 'acting out' an array of possible actions and results, participants became involved in a process that rehearsed political change and fostered critical engagement with alternative political models. Blurring the lines between actor and spectator, Boal's theatre encouraged citizens to 'give their opinions, discuss the issues, offer counter-arguments and share in the responsibility' for developing new local and national policy. In 1971 however, Boal's work drew the attention of the military dictatorship and he was arrested and tortured. After four months he was released and sent into exile, spending five years in Argentina before that country also descended into a climate of repression and violence.

Back in Brazil, graffiti again emerged as a popular form of expression for dissenting views. In the late 1960s, slogans including: '*Abaixa a dictadura*' (Down with the dictatorship); '*não à dictadura*' (no to dictatorship); '*não mais tortura*' (no more torture); '*militar fora*' (military out) appeared around urban zones, usually under the cover of darkness to preserve the anonymity and physical security of the producers. It was in this context too that artist, Cildo Meireles developed his so-called 'mobile graffiti'. In 1970, he anonymously inscribed the words 'Yankees go home!' on to empty Coca-Cola bottles that were returned and re-circulated by the company. This intervention, *Insertions into Ideological Circuits: Coca-Cola Project*, aimed at raising people's awareness of the 'ideological circuits' that surround them and penetrate into everyday life through basic acts of consumption such as drinking a soft drink. Meireles relinquished authorship over his 'mobile graffiti' as an act of basic pragmatism and self-preservation. Curator Kynaston McShine's commentary from 1970 reminds us of the risks that were associated with art-activism: 'If you are an artist in Brazil, you know of at least one friend who is being tortured' (see Maroja 2014).

Yet, '[s]treet art discourse was not always anti-regime' during the dictatorship period (Chaffee 1993: 134). The pseudo-democratic system engineered by the military regime featured regular elections where the two sanctioned parties could compete for votes. These elections tended to generate a broad range of street art as the ARENA and MDB vied for support with devices ranging from sophisticated lithograph posters to casually painted slogans and symbols. Other pro-system interventions sought to promote the regime's ideology of developmental nationalism. During the period 1969–1975, the Brazilian economy experienced a period of exceptional growth, dubbed by many observers as an 'economic miracle'. This period of growth coincided with the country's World Cup victory in 1970. Capitalising on these successes as an opportunity to bolster national unity and pride, the Medici government distributed posters featuring the footballer Pele scoring a goal, accompanied by the slogan '*Ninguem segura mais este pais*' (Nobody can stop this country now).[5]

Promoting growth through industrialisation, the Medici regime embarked on a range of large-scale industrial projects including the construction of the Trans-Amazonian Highway and the world's largest hydroelectric dam in the Rio Paraná, at Itaipu. These projects were lauded by the government and a subservient mainstream media as opportunities to lift the hinterland out of poverty. In practice, indigenous lands were annexed and decimated. The era of the 'economic miracle' doubled as the '*era dos desaparacidos*' (era of the disappeared), as hundreds of political dissidents, tribal peoples and others with supposedly leftist or anti-governmental affiliations were removed from society, tortured and killed by order of dominant military hardliners. Between 1964 and 1985, the military was responsible for over 400 documented deaths. It tortured[6] or exiled thousands more. Records have come to show that the highest number of human rights violations occurred during Medici's period in office, known popularly as the '*anos de chumbo*' or years of lead.

The Geisel period, which lasted from 1974 to 1979, saw an easing of repressive military actions and brought with it the very first steps in a long and protracted process of re-democratisation (Encarnación 2003). These democratising steps were to come in fits and starts, causing moments and resurgences of hope, creativity, disappointment and disenchantment amongst the broader populace. During this phase, the artist-activists Milton Sogabe, Eduardo Duar, Zé Carratu, César Teixeira, Jaime Prades, Rui Amaral and Carlos Delfino came together to form São Paulo's first documented street art crew or collective, *Grupo Tupinãodá*.

Prior to coming together as *Tupinãodá*, the artists had pursued somewhat different paths. Some were trained artists, while others were unschooled in the arts. Rui Amaral began painting on the streets as a part of the youth brigade of the *Partido dos Trabalhadores* (Workers Party or PT), while Ze Carratu described himself as the son of Italian anarchists and a student of 'Paris 68'. Jaime Prades had produced political cartoons and pamphlets for left-wing activists at the publishing house *Editora Abril* (Prades 2011a) before taking to the streets. Having all grown up during the dictatorship era, these young people all shared a grievance. Their opportunities had been formed and in many ways impeded following the imposition of the Institutional Acts. They had looked on as many of São Paulo's cultural spaces and historical buildings were demolished to make way for developers and financiers to move in (Carratu 1989). Moreover, violence and harassment directed at students, artists and workers in the São Paulo area had compounded their acrimony for the regime. Prades notes that, 'it was the time of the military regime, and in our generation you could not remain neutral. Our generation was militant, we took that to heart... and this manifested in the mobilisation of *Tupinãodá'* (Prades 2011a).

Carratu, Sogabe, Prades and Amaral started making individual, small-scale street art interventions across the neighbourhoods and districts of São Paulo. Following his spell painting for the PT, Amaral set about covering the streets and walls of the Vila Madelena neighbourhood with hundreds of tiny black stick figures. Amaral's stick figures were hand drawn; each one frozen in a unique gesticulation or dance move. In 1983, Milton Sogabe began producing figurative stencils in and around the neighbourhood of Pinheiros. Sogabe's crude stencils often depicted Brazil's tribal peoples and hunter-gatherer communities, armed with spears and arrows. Meanwhile, Carratu took to producing hasty chalk drawings, including abstract emblems and coloured patterns. He would sketch these over squares of wall that had been primed with black paint.

The artists would often exploit the cover of darkness or choose desolate areas such as abandoned recreation centres and demolition zones to produce their graffiti. As Carratu (1989) states, part of the rationale for this was rescuing public and cultural spaces. Years of repression directed at student organisations and social groups had led to a decline in creative local and community projects. Recreation centres lay abandoned and commentators describe the emergence of a *vazio cultural* (cultural void) (see Dunn 2001). However, another strategic imperative for painting in abandoned areas was the need to skirt the gaze and censorship of the authorities. Amaral was detained by the police in the early 80s and freed following an intervention by Amnesty International. He emphasises that

Figure 3.4 Stencil, 1983. Milton Sogabe.
Photograph Courtesy of Jaime Prades

Figure 3.5   Street art, circa 1983. Grupo Tupinãodá.
Photograph Courtesy of Jaime Prades

the context was a dangerous one: if caught painting oppositional messages, the street artists would have invited brutality from the *Departamento de Ordem Política e Social* (Department of Social and Political Order or DOPS). To self-attribute was not an option. To work quickly was a must.

Reflecting on the motivations for his early chalk graffiti, Carratu (1989) claimed that society was headed for a radical break and that the role of the artist was to help bring this about, by signalling '*chãos*' through art. Threatened by the elements from the very moment of production, Carratu's chalked images impressed a sense of ephemerality and the constant possibility of wiping away the existing with the new. Meanwhile, Amaral described his stick figures as 'part of an invented utopia', a non-real, ideal place where people might express themselves freely without harassment from the police (Amaral 2011). These expressions worked on different levels to offer cues for political change. For Amaral, there was an affective and indeed performative component at play: 'producing the figures gave to me some kind of release, they were freedom, presence... Art is determined by something more than what goes on in your head. It is also in your heart, you know?' (Amaral 2011). Further, in offering visions of *chãos* and utopia respectively, Amaral and Carratu's interventions served as departures from the norm, signalling that the world might be different somehow.

Yet, even if onlookers did not understand these gestures to other worlds, the presence of unsanctioned inscriptions across the city signalled a breach of the regime's machinery and capacity for total control. It is perhaps no surprise then that as Amaral (2011) reflects: '[the authorities] made an effort to cover up our graffiti.'

The decision to organise formally as a collective can be seen as another strategic move. As Prades (2011b) explains, 'We wanted to go bigger, bolder and reach more people'. The constant fear of attracting attention imposed limits on the size, style and detail of interventions that the artists could produce when working alone. Meanwhile, if they worked together, then one or more could keep a lookout while the rest painted. They began working alongside one another in the same spaces, which had a big impact on the style and scale of their creative outputs. Meeting regularly, they started discussing their ideas, blending their work together and mixing up their styles. Painting side by side, they were able to expand their coverage, which heightened the visibility and impact of their aesthetic disruptions.

> We regarded our collective work as a kind of 'strategic poetry'… I suppose we had the collective aim to create 'strange situations', turning our street art into a disruptive force. As our art evolved we would think about the functions and dynamics of the city, its velocity, its inhabitants and the relationship between the people and the public spaces.
>
> (Prades 2011a)

As the artists were often in contact with one another for hours at a time – painting, talking and socialising – they formed attachments to one another, to the spaces where they painted and to the ideas that they shared and discussed. A particularly strong influence over *Grupo Tupinãodá* were the Brazilian modernists and in particular, Oswald de Andrade's '*Manifesto Antropófago*' (Cannibal Manifesto). Having grown increasingly cynical of dominant narratives about Brazil's past, present and its future – particularly the vision of *ordem e progreso* (order and progress) pursued by the regime, the artists were keen to take up the modernists' call to reclaim and recast (or cannibalise) Brazilian national identity. Dunn (2001) offers a detailed exploration of Brazilian modernism, identifying two key imperatives behind the movement: the futurist and the primitivist. The futurist imperative called for 'formal experimentation, engagement with technology, and the representation of urbanity' (Dunn 2001: 14), while the primitivist element evoked the quotidian, or everyday cultural practices and experiences of the *povo* (the people or the masses).

The subject of identity became a strong organising theme for the artists of *Grupo Tupinãodá*. Their works addressed and explored aspects of both urban development and local culture, particularly the histories and traditions of Brazil's non-white population. Street art, itself an experimentation with form in the urban environment, represented one way of bridging and melding the rooted and more transformative ambitions of cultural cannibalism. Hence, Sogabe's tribal stencils, supplanted those traditional groups afflicted by the regime's developmental projects in the Amazon Basin to the centre of São Paulo, giving them a presence in the built environments generally upheld as symbols of modernity and progress. Meanwhile, Jaime Prades' stencil adaptation of Tarsila do Amaral's famed 'Uma Negra', sought to make the painting more accessible by bringing it to streets where it could be enjoyed by the 'excommunicated' segments of society:

the goal of my own work was to reproduce this highly regarded painting; to dislocate it from the institutions of high art and make it truly available to the mass audience by bringing it to the street. I suppose that in a way it was like a type of Brazilian 'pop art'.

(Prades 2011b)

The group shared a headquarters with the geographer Antonio Carlos Robert de Moraes in the Louis Anhaia alley, Vila Madalena. Their critical posture was

Figure 3.6 'Uma Negra' Stencil, circa 1982. Jaime Prades.
Photograph Courtesy of Jaime Prades

reflected in their choice of name, taken from a playful poem by de Moraes which drew on the question of identity in *modernismo*: '*Você é tupi daqui ou tupi de lá, e tupiniquim ou tupinãoda?*' (Are you Tupi from here? Or Tupi from there? Are you *Tupiniquim*, or *Tupinãodá*?). '*Tupiniquim*' was the language and name of the indigenous inhabitants of Brazil – reportedly cannibals – who made up the majority of the population when European colonisers arrived on the continent. In modern-day Brazil, the term 'Tupi' has often been used as a perjorative term for 'Brazilian national'. '*Tupinãodá*' is an invented and somewhat ambiguous term. Broken into composites, *Tupi não dá*, it denotes resistance and resilience: the particle *não* means 'no' or 'not' and the verb *dar* can mean 'to give' or 'to give in'.

Distensão: opening the pressure cooker

Over the course of the 1980s new opportunities for political expression opened up and the government experimented with heightened tolerance as part of the negotiated transition. By the end of 1984, the government party had splintered and a strong opposition coalition had formed around the idea of direct presidential elections. A three-month long campaign was initiated by the PT along with the *Partido del Movimiento Democrático Brasileño* (Democratic Movement Party or PMBD), *Partido Democrático Trabalhista* (Democratic Workers' Party or PDT), *Partido Comunista Brasileiro* (Brazilian Communist Party or PCB) and the *Partido da Social Democracia Brasileira* (Brazilian Social Democratic Party or PSDB) to pressure Congress into supporting a constitutional amendment allowing for direct elections.

In 1984, several million people participated in protests across Brazil and the slogan '*Diretas ja*' (Direct elections now) appeared on city walls, pavements and roadside embankments (Chaffee 1993). The movement for direct elections took up the yellow of the Brazilian flag (symbolising wealth) as its official colour. While no constitutional amendment initially resulted, the campaign had something of a transformative impact on society: 'It awakened the electorate, realigned political forces, split pro-government forces in Congress' (Chaffee 1993: 139). In this context, *Tupinãodá* also became tentatively more bold in their collective projects. Over the course of several months, the group collected a vast number of full rubbish bags coloured black, yellow and blue. They fashioned these bags on an expansive rectangular green lawn at the University of São Paulo in a geometrical composition that recalled the *Bandeira do Brasil* (the Brazilian flag). Years later, Jaime Prades explained that through this intervention, the group wanted to denounce the regime's monopoly on the Brazilian national identity:

> The dictatorship had appropriated the colours of the flag in a negative manner: in the ways it used football as a political tool; in its encouragement of damaging commercial practices; and, in its repressive measures against the Brazilian people themselves... We felt a need to express ourselves, express our frustrations. We, as Brazilians, wanted to bring alive elements from our own imaginations, not to repeat the codes handed down to us by those in

Figure 3.7   'Bandeira do Brasil', circa 1984. Grupo Tupinãodá.
Photograph Courtesy of Jaime Prades

power. We wanted to give new interpretations to objects and symbols that people see everyday.

(Prades 2011b)

In spite of a vicious poster campaign sponsored by the military, civilian candidate Tancredo Neves was successfully elected to the Presidency in 1985. Many took to the streets to celebrate the democratic restoration. However, before he was inaugurated by ceremony, Neves fell ill and died of a heart attack, leaving Vice-President Jose Sarney, to gain office by default. Sarney instituted some liberalising measures with immediacy, legalising all political parties, restoring the constitutional promise of direct presidential and mayoral elections and enfranchising illiterate members of the populace (Skidmore 1988). However, Sarney also had long-standing connections to the outgoing regime, having previously led the government-sponsored party, ARENA. Atencio (2014) opines that Brazil's first civilian president in 21 years took no meaningful steps to hold the officials of the former military government to account for their crimes. Strikingly, even in the wake of the 1985 *Nunca Mais* Report, which compiled thousands of classified documents detailing the use of torture during Médici's government, a 1979 law guaranteeing amnesty was upheld (Skidmore 1988). After 1985, democracy remained tarnished by the persisting presence of military personalities within the new administration, which was not the product of the people's vote but of a democratically elected yet partial constituent assembly (Encarnación 2003). Skidmore (1988) estimates that retired officers occupied an estimated 8–10,000 top posts in the 'new' government and the state enterprises. Moreover, 'there was still the lavishly funded SNI,[7] operating in the shadows and heavily staffed with military officers' (ibid.: 273). As a result, key authoritarian measures were left in place, such as the Press Law, the National Security Law, and Decree No. 1077, which banned the transmission of unapproved live broadcasts and publications (Green and Karolides 2009). Faced with these less-than-democratic continuities, public hope for accountability and proper implementation of the rule of law faded and discontent emerged in its wake.

From 1986, visual demands for Sarney to step aside began to decorate the streets around São Paulo and disenchanted *paulistas* continued to mobilise around the demand for '*Diretas ja*'. The PT produced posters demanding '*Fora Sarney*' (Sarney out) and other documented slogans included '*indiretos eleições – nunca mais*' (indirect elections – never again'); '*Morte a Sarney: bandido e traidor*' (Death to Sarney – crook and traitor) and '*Sarney é um vigarista*' (Sarney is a crook). During this period, *Grupo Tupinãodá* continued their street art interventions with even greater zeal, producing an array of large-scale paintings around the city of São Paulo. The group increasingly worked in the light of day and tackled expansive public spots with less regard for the authorities. Perhaps most notably, *Tupinãodá* were the first to paint in the now famous *beco do Batman* (Batman alleyway) as well as the large concrete tunnel running beneath the Avenida Paulista in São Paulo's financial centre. As Magalhães (2009) describes, these new works were playful in terms of the imaginary, invoking bold colours, patterns

Figure 3.8   Mega-graffiti, circa 1986. Grupo Tupinãodá.
Photograph Courtesy of Jaime Prades

and futuristic emblems such as rockets and robots. In terms of both style and scale, these works pre-empted contemporary *graphite*. But the influence of the modernists is also arguably present in the experimental melding of futuristic imagery and dark, cavernous urban spaces.

Halsey and Young (2006) comment on the strong affective dimension of street art production, arguing that street art or graffiti can offer new sites and images through which bodies can connect with themselves and with the world. Drawing on the work of Brian Massumi they claim that, with intensified affect, comes a stronger sense of embeddedness in a larger field of life – and quite possibly, a heightened sense of belonging. Commentaries from *Tupinãodá* seem to support these assertions. When asked about the shift in the group's style and fervour, Prades suggests that initially the re-instatement of certain rights and freedoms accompanying the political opening gave the group a greater confidence to paint in the open, but he adds: 'What I think is that when the pressure cooker started to open, paint gushed out!' (Prades 2011a). Even as the prospects for substantive political change appeared to diminish with Sarney's limited programme of reform, it became impossible to contain this energy (ibid.), which drove them to paint and brought more and more people onto the streets and into contact with their interventions. As Ze Carratu says, 'The power to act in the street, to occupy the walls, abandoned buildings and locations with weird architecture, is also an important kind of force.' Reflecting, Prades also articulates that the group's mega-graffiti helped 'to show people what is possible' (Prades 2011b). Certainly, *Tupinãodá's* interventions in the Paulista and Reboucas tunnels were soon followed by a spate of other aesthetic interventions across the city, as well as invitations to appear in gallery shows. Art critic Barreira describes 1987 as the year in which *graffiteiros* descended on the city of São Paulo, marking a fundamental shift in the balance of threats and opportunities for street artists (Barreira 1987). Of course, in most existing commentaries, it is here that the story of Brazilian street art begins.

Tupinãodá's mega-graffiti in the Avenida Paulista tunnels is remembered today by a cross-section of São Paulo's residents as well as many contemporary street artists, including the internationally renowned *Os Gemeos* (see Art Crimes 2000). Part of the reason for this is that *Tupinãodá's* interventions broke with the urban visuality in a profound and unexpected way. The military regime's attempts to project its own aesthetic ideals onto the public space, coupled with its practices of censorship, had resulted in a lull of civic enterprise and public activity. *Tupinãodá's* large-scale graffiti set a visual precedent for change, encouraging new modes of public expression that contributed to the re-awakening of a fractured and demoralised civil society.

Brazilian street art: a landscape in motion

In spite of their impact and acuity, *Tupinãodá* – like other dictatorship era artists – have until recently received very little coverage from major galleries, museums and other cultural institutions in Brazil. This has much to do with the country's

Figure 3.9   Mega-graffiti, circa 1986. Grupo Tupinãodá.
Photograph Courtesy of Jaime Prades

rather particular route to democracy, involving a protracted transition through which many former military personnel maintained positions of power and a 'feigned amnesia – combined with overt pride in the authoritarian past' (Pereira 2005: 162). This confluence of factors significantly limited the scope for public debate, cultural activity and legal action pertaining to the violence of the dictatorship years (Ryan 2016). Hence, during the 1990s, acts of memorialisation through street art seem to have been almost striking in their absence. Meanwhile, a 'second generation' of internationally renowned artists including *Speto, Os Gemeos, Viche* and others have preferred to trade in the circulation of cheerful imaginaries, colourful characters and vectors; turning Brazilian street art into a 'happy object' as Sara Ahmed might say.

Ahmed's writing traces how certain objects become imbued with positive affects and are treated as desirable items, perceived as necessary for a good life. Drawing on the work of John Locke, she argues that we judge something to be good or bad according to how it affects us, and this experience stays with us, informing future encounters with said object and creating the *promise* of happiness. Yet, Ahmed contends that 'we need to question what is appealing in the appeal to happiness and good feeling' and to ask what is lost or overshadowed in the move to fulfil promises of happiness. She writes, 'What concerns me is how much this affirmative turn actually depends on the very distinction between good and bad feelings that presumes that bad feelings are backward and conservative and good feelings are forward' and progressive' (ibid. 42).

This distinction has been evidenced in Brazil's 'happy street art' and it has had a determinate political and economic role. The colourful and playful graphic works of the 1990s and early 2000s mirrored the optimism of the time; the idea that a newly democratic Brazil could make its advance, shedding the vestiges of military repression with its people looking steadfastly onward and upward. Moreover, not only has there been a movement to sell ideas and products – from beer and skateboards to global sports mega events – through this bright and uplifting medium, street art has also become a tool for the 'regeneration' of Brazil's *favelas* or slums. Local and international street artists and community groups have coordinated a range of innovative projects with marginalised youth in the favelas, teaching discipline, nonviolence and vocational skills through painting. For some, these creative programmes have been truly transformative. As Marcos Rodrigo Neves indicates in an interview with *Smithsonian Magazine*, street art saved him from gangs and drugs, offering an opportunity for upward social mobility where no others existed (Hammer 2013).

However, recognising the allure of street art for international tourists, municipal and federal authorities have also become increasingly involved in the coordination, support or financing of street art initiatives in the favelas. In March 2009, the Federal government passed law 706/07 which effectively decriminalises street art. Cities including São Paulo now integrate street art into urban policies. Geographers such as David Ley (2003) have raised valid concerns about the symbiosis of art and symbolic violence in these kinds of contexts however, outlining how the arrival of artists can unwittingly usher in a process by which objects move from 'junk to art and then on to commodity'. The particular concern here is about the *aesthetisation* of the favela through 'happy street art': a process through which the real social problems and needs of local residents become cemented, homogenised and overshadowed by municipal authorities and tour operators who beautify and sensationalise urban poverty in pursuit of economic rent.

In the complex and evolving Brazilian landscape, however, other processes of change and resistance are also afoot. Since 2005, the Federal Government, under the leadership of the *Partido dos Trabalhadores*, actively promoted a new culture of memory around the dictatorship years, initiating a range of truth and memory projects – including *caravanas*, art installations and most prominently the Comissão Nacional da Verdade (National Trust Commission or CNV). These endeavours have opened up discursive space, challenging social amnesia and bringing entrenched structures of violence such as police impunity, racism and environmental degradation to the forefront of public consciousness (Ryan 2016). In this context, a venerable 'third generation' of street artists is emerging, using styles and scales pioneered by *Tupinãodá* to give renewed visibility to marginalised actors and issues. São Paulo based artists, Paulo Ito and Fabio de Oliveira Parnaiba, better known as Cranio (or 'Skull' in English) form a part of this new movement.

Paulo Ito's political 'memes' seek to draw attention to a broad range of social issues that affect modern Brazil. One of his paintings, completed in Fortaleza – the fifth largest municipality of Brazil and one of the country's most visited touristic capitals – explores the issue of sexual tourism and in particular the abuse of minors. Although exact figures remain elusive, it is well known that the sexual exploitation of children is a big problem in Brazil, and one expected to be compounded by global sports mega events of 2014 and 2016 (Defense for Children International 2013). Ito addressed this issue in a 2014 composition which showed an older white male – presumably a tourist – presenting a doll to a young girl, all the while hiding a rose behind his back. Meanwhile, Ito's work achieved global recognition in 2014 when he depicted a hungry boy being offered a soccer ball instead of food on the wall of a schoolhouse in São Paulo. The image was shared widely across the internet, becoming a visual adjunct to unfolding protests against the World Cup. Ito told *Slate Magazine* that the football painting had been his most powerful creation, highlighting how it tapped into and amplified the circulating public sentiment. 'People already have the feeling and that image condensed this feeling', Ito told Stahl (2014).

Cranio's blue tribal characters meanwhile seek to draw attention to the issues of cultural assimilation and the negative impacts of consumerist culture on Brazil's indigenous communities. In a 2014 interview with Salim Jawad, Cranio expands on his work:

> I paint indigenous people, but I like to transform them to adapt any kind of culture or situation… My main objective is to show people the reality of Brasil [by] painting the many issues happening right now in the country. Usually they are environment, politics, corruption and consumerism.

[My characters] are absorbing the occidental culture and that is why they are blue. They all got sick [from] this life far away from the forests, going to the big cities, trying to sell their land in exchange for consumerism... to buy superfluous gadget[s] like an iphone, sneakers, etc.

Summary

In Brazil, street art forms – from posters and murals to *pichação* and mega-graffiti – have been mobilised by a broad spectrum of political actors to project ideas, make claims and express un-coded sentiments. Street art has accompanied periods of political change, such as elections and *coup d'etats*, as well as periods of representative crisis and democratic change.

The first half of this chapter focused on early street poster art in Brazil, underlining how posters were used to frame issues, mobilise public support and generate resources for vying factions during the Paulista War. Drawing on the reflections of Susan Tschabrun (2003) and others, it suggested that there are two key ingredients for an impactful political poster: i) a strong symbiosis of word and image, and ii) an ability to rouse the viewer by modulating affect. Moreover, it argued that street poster art serves a useful evidentiary function, providing insights into the particular configurations and types of power circulating at a particular historical moment.

In the second half of the chapter, coverage moved towards the birth and evolution of *graphite*, first exploring the work of *Grupo Tupinãodá*. Most of *Tupinãodá's* street art interventions took place against the protracted transition from authoritarianism to formal democratic rule. The shifting balance of threats and opportunities over this period was reflected partly in changes to the style, sophistication and willingness of the *Tupinãodá* artist-activists to self-attribute their work. Thus, when the risk of retaliation was at its highest, interventions were anonymous and speedily produced in order to skirt the gaze and censorship of the authorities. Later, collective production came as a strategic move driven by the desire to produce larger and more complicated pieces without police incursion. Following the transition to formal democracy, *Tupinãodá's* use of street art evolved once again, reflecting the evolving stakes and the changing sentiments wrought by the *distensão* (decompression/opening) and its discontents.

However, *Tupinãodá's* work did not just respond to changes in the political environment or 'opportunity structure', it actively resisted, intervened and transformed it in various ways. *Tupinãodá's* street art revealed alternative ways and options for engaging politically under authoritarianism. Amaral's imagined utopias and Carratu's pre-emptions of chaos proposed and maintained that the world could be otherwise. As Duncombe (2012) and others maintain, the power of utopian imagery lies not in the specifics of future visions themselves, which can have the effect of closing down on imagination, but rather in the fact that such imagination is possible in the first place.

From *Tupinãodá's anthropofagia* to the work of second and third generation street artists, it is possible to see how street art has consistently provided a medium through which to represent and contest what it means to be 'Brazilian'. Furthermore, through the artists' testimonies we can also get a glimpse of the affective interplay that occurs around and through street art. Whether capturing and amplifying the public mood or offering an outlet for personal feelings, political street art in Brazil exemplifies Jill Bennett's claim that affect is the natural medium of practical aesthetics.

Notes

1 Much of today's street art is described as *graphite*, a style distinctive in its boldness, colour and scale. Brazilian graphite has become commonly associated with themes such as celebration, happiness and progress.
2 São Paulo's economy was driven by coffee, and Minas Gerais was Brazil's largest dairy producer, hence the name 'coffee with milk' politics.
3 *Folhetos* are a form of bulletin or pamphlet, generally distributed on the street.
4 Today the term *pixação* is used to denote a unique style of cryptic tagging which gradually colonised the façades and tops of buildings in the capital over the course of the 1990s. Chastanet (2005) writes that contemporary *pixação* is unique because the pixadores have developed a new imaginary calligraphy, influenced by the heavy metal and hardcore logos of record sleeves of the 1980s. He continues that *pixação* functions as a 'parallel prestige economy' organised by writing, where the act of writing one's name and performing one's signature in a public space is more about seeing than reading.
5 For more details on the regime's political use of the World Cup, see Levine (1980) and Flynn (1973).
6 In an article for Al Jazeera, Picq (2013) translates the testimony of filmmaker Lucia Murat who was tortured in her early twenties. During the Medici era, Murat was arrested and subjected to 'intense beatings followed by sessions of electroshocks while hanging naked and wet from a metal bar. Her naked body was covered with cockroaches, and for months, she spent nights tied up enduring what her interrogators called "sexual scientific torture"'.
7 The *Serviço Nacional de Informações* (National Information Service or SNI) was the Brazilian intelligence agency established under the Castelo Branco government. Originally a civilian organization, it came under the leadership of military hardliners and emerged as a leading force in the hunt for communists.

References

Ahmed, S. (2003) Affect Economies. *Social Text* 79. 22(2), pp.117–139.

Ahmed, S. (2004) *The Cultural Politics of Emotion*. New York: Routledge.

Ahmed, S. (2010) Happy Objects, *in* M. Gregg and J. Seigworth (eds) *The Affect Theory Reader*. London: Duke University Press, pp.29–51.

Amaral, R. (2011). Interviewed by Ryan, H.E. São Paulo/Brazil, 28 September 2011.

Art Crimes (2000) *Interview with Os Gemeos*. Available from: www.graffiti.org/osgemeos/osgemeos_2.html

Atencio, R. (2014) *Memory's Turn: Reckoning with dictatorship in Brazil*. Madison: University of Wisconsin Press.

Barreira, W. (1987) The Colors of the Streets. *Veja Magazine*. Retrieved from: www.jaimeprades.art.br/?area=5&sec=1&item=2&lang=_en

Carratu, Z. (1989) Interviewed by Carlsson, C. and Manning for the documentary 'Brazilian Dreams', C. São Paulo/Brazil, 1989.

Chaffee, L.G. (1993) *Political Protest and Street Art: Popular tools for democratization in Hispanic countries.* Westport, Connecticut: Greenwood Publishing Group.

Chastanet, F. (2005) The architecture of São Paulo, Brazil, is covered by a unique form of calligraphic graffiti. Eye. 56. Retrieved from: www.eyemagazine.com/feature/article/pichao-extract

Chasteen, J.C. (2006) *Born in Blood and Fire. A Concise History of Latin America.* 2nd edn. London: W.W. Norton.

Chong, D. and Druckman, J.N. (2013) Counterframing effects. *The Journal of Politics.* 75(1), pp.1–16.

Davila, J. (2006) Myth and Memory: Getúlio Vargas's Long Shadow over Brazilian History, *in* Hentschke J.R. (ed.) *Vargas and Brazil: New Perspectives.* New York: Palgrave Macmillan, pp.257–282.

Defense for Children International (2013) *Review of Brazil.* Retrieved from: www.defenceforchildren.org/search/brazil

Duncombe, S. (2012) Imagining No-place. *Transformative Works and Cultures,* 10, doi:10.3983/twc.2012.0350

Dunn, C. (2001) *Brutality Garden: Tropicália and the Emergence of a Brazilian Counterculture.* Chapel Hill, North Carolina: University of North Carolina Press.

Dunn, C. (2014) Desbunde and Its Discontents: Counterculture and Authoritarian Modernization in Brazil, 1968–1974. *The Americas.* 70(3), pp.429–458.

Encarnación, O.G. (2003). *The Myth of Civil Society.* New York: Palgrave Macmillan.

Fleischer, D. (2005) Brazil: From Military Regime to a Workers' Party Government, *in* Knippers-Black, J. (ed.) *Latin America: Its Problems and Its Promise.* Massachusetts: Westview Press, pp.470–500.

Flynn, P. (1973) The Brazilian development model: the political dimension. *World Today.* 29(11), pp.481–494.

Goffman, E. (1974) *Frame Analysis: An essay on the organization of experience.* London: Northeastern University Press.

Green, J. and Karolides, N. (2009) *Encyclopedia of Censorship.* 3rd edn. New York: Facts on File Inc.

Green, J. and Karolides, N. (2014) *Encyclopedia of Censorship.* New York: Facts on File.

Halsey, M. and Young, A. (2006) Our desires are ungovernable: Writing graffiti in urban space. *Theoretical Criminology.* 10(3), pp.275–306.

Hammer, J. (2013) A Look Into Brazil's Makeover of Rio's Slums. *Smithsonian Magazine.* Retrieved from: www.smithsonianmag.com/people-places/a-look-into-brazils-make over-of-rios-slums-165624916/?no-ist

Hilton, S.E. (1979) Brazilian Diplomacy and the Washington-Rio de Janeiro 'Axis' during the World War II Era. *The Hispanic American Historical Review.* 59(2), pp.201–231.

Huffington Post (2014) *26 Best Cities in the World to See Street Art.* Available from: www.huffingtonpost.com/2014/04/17/best-street-art-cities_n_5155653.html

Jasper, J.M. (1998) The emotions of protest: Affective and reactive emotions in and around social movements. *Sociological Forum.* 13(3), pp.397–424.

Koonings, K. and Kruijt, D. (1999). *Societies of Fear: The legacy of civil war, violence and terror in Latin America.* London: Zed Books.

Lauerhass Jr, L. (1979) Who Was Getúlio? Theme and Variations in Brazilian Political Lore. *Journal of Latin American Lore.* 5(2), pp.273–290.

Levine, R. (1980) The Burden of Success: 'Futebol' and Brazilia Society through the 1970s. *Journal of Popular Culture*. 14(3), pp.453–464.

Levine, R.M. (1998) *Father of the Poor?: Vargas and his Era*. Cambridge: Cambridge University Press.

Ley, D. (2003) Artists, aestheticization and the field of gentrification. *Urban Studies*, 40(12), pp.2527–2544.

Magalhães, F. (2009) Jaime Prades: Art in Context, *in* Prades, J. (ed.) *A Arte de Jaime Prades*. São Paulo: Olhares.

Mainwaring, S. (1986) The Transition to Democracy in Brazil. *Journal of Interamerican Studies and World Affairs*. 28(1), pp.149–179.

Manco, T., Lost Art and Neelon, C. (2005) *Graffiti Brasil*. London: Thames and Hudson.

Maroja, C. (2014) Transforming Spectators into Witnesses. Sztuka i Dokumentacja, nr 10. Available from: www.journal.doc.art.pl/pdf10/art_and_text_camila_maroja.pdf

Morris, A. (2000) Reflections on social movement theory: Criticisms and proposals. *Contemporary Sociology*. 29(3), pp.445–454.

Nelson, C. and Rutstein, J.S. (1995) Posters as a library resource: the international poster collection at Colorado State University. *Collection Management*. 20(1–2), pp.115–138.

Pereira, A. (2005) *Political (In)Justice. Authoritarianism and the Rule of Law in Brazil, Chile, and Argentina*. Pittsburgh: University of Pittsburgh Press.

Picq, M. (2013) Rescuing the memory of Latin America. *Al Jazeera*. Retrieved from: www.aljazeera.com/indepth/opinion/2013/06/20136611144173414.html

Prades, J. (2009) *A Arte de Jaime Prades*. São Paulo: Olhares.

Prades, J. (2011a) Interviewed by Ryan, H.E. São Paulo/Brazil, 29 September 2011.

Prades, J. (2011b) Interviewed by Ryan, H.E. São Paulo/Brazil, 30 November 2011.

Rickards, M. (1971) *The Rise and Fall of the Poster*. London: David and Charles.

Ryan, H.E. (2016) From Absent to Present Pasts: Civil Society, Democracy and the Shifting Place of Memory in Brazil. *Journal of Civil Society*. doi:10.1080/17448689.2016.1165 484

Seidman, S.A. (2008) *Posters, Propaganda, and Persuasion in Election Campaigns Around the World and Through History*. New York: Peter Lang Publishing.

Skidmore, T.E. (1988) *The Politics of Military Rule in Brazil 1964-85*. Oxford: Oxford University Press.

Tarrow, S. (1998) *Power in Movement*. 2nd edn. New York: Cambridge University Press

Tschabrun, S. (2003) Off the Wall and into a Drawer: Managing a Research Collection of Political Posters. *The American Archivist*. 66, pp.303–324.

Zald, M.N. (1996) Culture, ideology, and strategic framing, *in* McAdam, D., McCarthy, J. and Zald, M. (eds) *Comparative Perspectives on Social Movements: Political opportunities, mobilizing structures, and cultural framings*. Cambridge: Cambridge University Press, pp.261–274.

4   *Pintadas* and performances

Street art, identity and resistance in Bolivia

Descending the hill from the world's highest airport at El Alto into the valley that contains Bolivia's capital city of La Paz, one encounters a dazzling array of political commentary. Concrete walls that spiral downwards to line the road and the facades of once-white buildings that cleave into the reddish rocky landscape are covered with graffiti and wall painting. Some slogans are scrawled with little consideration to legibility or visual appeal. Others are obviously more considered, featuring large, bold lettering, decipherable from metres away. These interruptions to the surrounding landscape feature the use of spray paint, masonry paint and even tar to deliver messages of hope, resistance and loyalty. Captions have included '*Afuera USAID*' (Out USAID) '*Todo va a cambiar*' (Everything is going to change), '*Lucho por La Paz*' (Fight for La Paz), '*Te Amo MAS*' (I love you MAS).[1] On entering the wealthier Achumani area of La Paz one may even encounter large-scale stencils, wheatpastes and sophisticated murals. But La Paz and indeed Bolivia as a whole is little known for its street art production. Maximilliano Ruiz's book '*Nuevo Mundo*' documents street art from across Latin America, offering visuals from 13 different states. Yet, it fails to consider Bolivia among these (see Ruiz 2011). So, just *who produces street art in Bolivia? What does it achieve? And how?*

Bolivia is a country characterised by many paradoxes. While it is one of the richest countries in Latin America in terms of its natural resources, economically it remains one of the poorest countries in the region, harbouring stark inequalities and one of the lowest indices for per capita income. Bolivia is the world's second largest producer of tin and it is endowed with large deposits of bauxite, iron, zinc, gold and other highly valued minerals. Over the period of colonial rule, the extraction of over 62,000 metric tonnes of silver from mines at the *Cerro de Potosí* (Mount Potosí) was used to finance the exorbitant lifestyles of the Spanish aristocracy and pay off Spain's debts to its neighbours. For a time, Europe was awash with silver from Potosí (Branford 2004). However, colonial powers were more interested in extracting and exporting riches than developing local infrastructures. The war for independence left Potosí in ruins and paralysed its silver mining industry. Today, the only traces of Potosí's former glory are the *Casa de la Moneda* (The Mint) and a few religious artefacts from the Baroque era (Nash 1993).

Today, Bolivia is populated by over 10.6 million people, out of which a strong majority (over 60 per cent of the population) identify as 'indigenous'. The ILO confirms that Bolivia has the highest percentage of indigenous peoples in all of Latin America. Of this percentage, it is estimated that the majority are Quechua (50.3 per cent) and Aymara (39.8 per cent). Lowland peoples such as the Chiquitano (3.6 per cent) and Guaraní (2.5 per cent) are smaller in number. The departments of La Paz, Cochabamba, Potosí, Oruro and Chuquisaca have the highest indigenous concentration (ILO 2015). For centuries, Bolivia's indigenous Indian population was largely excluded from the political process, official culture and the benefits derived from the extraction and export of natural resources. As a result, there have been a great many fierce political struggles throughout Bolivian history, leading James Dunkerley (1984) to question whether the country has *'Rebellion in the Veins'*. One of the earliest recorded episodes of 'everyday resistance' in the territory of Upper Peru (now Bolivia) involved the spread of traditional dance as reaffirmation of Quechua culture. *Taki unkuy* literally translates as 'dancing disease' and refers to the eruption of dance in the streets and towns – an act of cultural and political resistance which culminated in a revolt against the Spanish colonisers in the sixteenth century.

Around 200 years later, the Túpac Amaru, Catari and Túpac Catari rebellions swept across the Southern Andes, 'threatening Spanish Control of its Latin American territories more than a quarter century before the Wars of Independence (1808–1825)' (Walker 2008: xxiii). José Gabriel Condorcanqui Noguera, a Spanish-speaking Quechua Indian who worked as a *cacique* or mediator between the colonial state and the indigenous people became increasingly disgruntled at the harsh labour conditions and heavily taxes imposed on his people. José Gabriel claimed descent from the 'last of the Incas' – Túpac Amaru – and took on his name, capturing and executing one of the most abusive *corregidores de indios* (overseers of the Indians),[2] before leading an uprising of approximately 60,000 Indians against the Spanish in 1781. In the same year, Túpac Catari and his (mainly) Aymara followers laid siege to the city of La Paz for a period of six months.

Although both movements were eventually supressed by the colonial state, Nash (1993: 1) claims that '[Bolivia's] history of struggle since the uprising of Túpac Amaru and his ally Túpac Catari against the Spaniards in 1781, to the advent of the Che Guevara *guerilla* movement in 1967, reveals its people to be among the most politically responsive' in the Latin American region. In spite of this, it was not until 1952 that Bolivia formally enfranchised Indians by granting universal suffrage and doing away with the *latifundio*[3] system. Bolivia's ratification of ILO Convention No. 169 in 1991 paved the way for greater recognition for indigenous groups and in 1994, the Constitution acknowledged the 'multiethnic and multicultural' nature of the Republic. Constitutional reforms in 2004 recognised indigenous peoples' rights to present political candidates directly. However, it was only in 2006, more than 500 years after the arrival of the Spanish, that Bolivians came to the ballot to elect an indigenous president.

In order make sense of political street art in Bolivia – to understand its evolution as a mode of protest, pedagogy and political expression – it is necessary to examine its development alongside key events and processes in the country's history. Hence, this chapter offers a journey through a long twentieth century, taking in the protracted and arduous struggle for substantive social, cultural and political equality for the country's indigenous population. The first part of this chapter provides a fairly detailed overview of the ways in which political street art developed and evolved in the Bolivian example up until the time of the 1952 National Revolution. The chapter moves on to explore the use of street art by the *Círculo 70* who operated in a context characterised by repression and brutality under the military dictatorship of the 1970s. It then goes on to discuss the various interventions of the anarcha-feminist movement *Mujeres Creando* and *Insurgencia Comunitaria*, both of which emerged after the democratic transition, disillusioned with the lack of progress in the protection of rights and political representation for marginalised groups. Across these cases, the chapter illustrates the power and importance of political street art in articulating alternatives and bringing marginalised actors and power asymmetries into view.

Affirmation for Indians? From Federal War to the Escuela-Ayllu Warisata

The area formerly known as Upper Peru gained its independence from the Spanish in 1825, following 16 years of struggle between discontent *criollo*[4] elites and the Spanish Crown. Simon Bolivar, the aristocrat, intellectual and revolutionary leader who liberated no less than six modern Latin American states, sat as the first president of the new independent Bolivian Republic. Half a century on however, Bolivia lost its coastal territory in the Atacama to Chile and became landlocked. After Bolivia gained its independence there was a short republican period dominated by the criollo elite, during which time a tribute system was maintained. This system required Indian communities to pay a tax in return for some level of autonomy *vis a vis* the state. Lucero (2008) highlights that revenue generated from the tribute system accounted for up to 75 per cent of all taxes collected in the new republic. Arandia (2011) explains that during this time, little or no attention was paid to the cultures and customs of indigenous population or members of the lower classes. The arts and literature in the new Bolivia largely ignored this broad swathe of the populace, marginalising its aesthetic and linguistic practices and ignoring its social and economic needs. As Eduardo Galeano (1974: 46) reminds readers, 'Masters of Indian pongos – domestic servants – were still offering them for hire in La Paz newspapers at the beginning of our century'.

The Federal War of 1898 saw Indians used as political pawns by the Liberal Party, who promised to defend communal landholdings in return for support. However, in the aftermath of the war the victorious Liberal leader Jose Manuel Pando failed to deliver on these promises, evoking a series of violent uprisings by *campesino* (indigenous peasant) groups. One of these uprisings, led by the Aymara colonel Pablo Zarate Willka, is noted by scholars as perhaps the largest Indian rebellion in Bolivian history. Moreover, as Balderston and Schwarz (2002: 171) put it, 'this rebellion reawakened in urban society, a fear of the Indian which had always been present since colonial times and unleashed rhetorical and physical abuse against him in the early part of the century, a period baptised by the historian Daniel Demelas as that of 'creole Darwinism'.' For some writers and artists of the time, the solution to this political upheaval was clear: it was necessary to carve indigenous culture more squarely into the national imaginary. As a part of this '*indigendist*' movement, made up of mostly an urban middle and upper class, artists began to paint indigenous subjects, often depicting them as mystical, alluring and timeless figures. In literature, writers such as Alcides Arguedas took the reader on visits to the countryside in encounters with supposedly 'authentic', autochthonous figures.

However, in attempting to portray a more realistic impression of Bolivian culture, these cultural producers often succeeded only in refashioning it according to their own, quite narrow, worldviews. Balderston and Schwarz (2002: 172) explain how, in the case of Arguedas' writing (see '*Wuata Wara*' and '*La Raza de Bronce*'), we find a narrative formula that combines a 'late nineteenth century European racialist thinking to understand the Indian and scientifically legitimise urban prejudices against him' with 'a revelation of the unjust conditions of indigenous life and work in the haciendas of the Bolivian highlands'. Quite apart from enhancing the affinity and sameness of peoples within one nation, the result was a set of representations that reinforced the idea of the Indian as 'Other' by underlining his supposedly pre-modern and pre-rational predilection. As Jen Webb (2009) reminds us, practices of representation do not just give presence to things and peoples, they also become productive of what we know and feel about them. Hence, the visual and literary representations of Indian peoples seen in the works of early '*indigenists*' not only incorporated Indian subjects as legitimate cultural matter for the first time, they also framed wider discourse and societal attitudes. As such, Indians were to be seen as a part of the nation, but they were not to be viewed on an equal footing with criollo and dominant white classes.

Between 1932 and 1935, Bolivia fought a bitter war against Paraguay over ownership of the Chaco, an expanse of semi-arid land that had sparked the interest of oil exploration companies. The Bolivian government drafted thousands of Indians to support the struggling army in their efforts to defeat Paraguay. In an attempt to guarantee unity and commitment from the conscripts, criollo military officers emphasised, '[The] Indian's duties and obligations, and incidentally, his rights, his citizenship, his equality before the fatherland' (Patch 1961: 126). For many Indians, the promise of constitutional rights and substantive equality was initially quite compelling. However, for the tens of thousands who were dumped into the Chaco wilderness, the reality of ethnic segregation loomed large. The higher ranking officers were invariably of European descent and often could not even communicate with their own soldiers, who were mostly untrained and largely illiterate Indians from Aymara or Quechua communities (Scheina 2003).

Not at all remiss of the bloodshed in the Chaco, others decided to take advantage of the new discourse of 'inclusivity' in order to pursue social policies, projects and investments that advanced the position of Indian peoples. In this context, Avelino Sinani and Elizardo Pérez founded the Warisata Ayllu School, an intercultural experiment in indigenous schooling that flourished on the high plateau (*altiplano*) between 1931 and 1940 (Larson 2011). In Aymara, '*wari*' means 'inner strength' and '*sata*' means 'to sow or plant'. With their plans for the school, Sinani and Pérez sought to develop an innovative new curriculum that would respect communal lifestyles and help young Indians to gain self-sufficiency and political confidence. Aymara communal work parties (known as *minkas*) donated lands in exchange for their education of their children and they collaborated with teachers to bake adobe bricks which were used to erect the school building. The school had fields, orchards and boats that were used to provision the school children. An Aymara council of *amawt'as* (wise elders) oversaw the issues of governance and discipline (ibid. 65).

The school eventually drew hundreds of students, and its reputation grew across Bolivia and the wider Andean region. Among its supporters, the ayllu school was acclaimed for providing a civic and cultural education that ran counter to the *mestizaje* ideology, which formed part of a broader scheme of cultural assimilation by political elites (Telles and Garcia 2013). Although they are sometimes overlooked in historical accounts, murals played a considerable role in the Warisata project of education and resistance.

In 1934, Aymara printmaker Alejandro Mario Yllanes painted a series of tempera murals on the schoolhouse walls at Warisata. Although the series was never completed, the murals were composed to portray and celebrate the daily labours and achievements of the community. Some of the murals depicted Aymara farmers, ferrymen and leather workers. Others portrayed images from Andean popular history and storytelling. Thematically, the murals corresponded with the school's ethos of *ayni*. In traditional Andean culture, the principle of *ayni* describes the complementarity of lifeforms; the ways that all things in the universe fit together through relationships of mutual aid and reciprocity. Hence, Yllanes' farmers and fisherman took only what they needed from the earth. They did not plunder, raze or destroy. Their visual presence called on students to uphold the same principles of sustainable living.

Carlos Salazar Mostajo, a writer, painter and former teacher at Warisata, reviewed much of Yllanes' work and opined that his murals were special not just because they took the working Indian as their subject but because, by extension, they paved the way for the contemplation of other groups in Bolivian society that had been marginalised by the political and cultural institutions of the day (Salazar Mostajo 1989). Moreover, as contrasted with the exoticising representative practices common to earlier *indigenist* cultural productions, Yllanes' works offered a view from inside the community. Perhaps for this reason his subjects had a certain sense of the 'everyday' about them. They were not sat poised and passive, waiting for an audience to examine them. Rather, they were depicted in action, going about their daily chores – fishing, tanning etc. with a dynamism and strength. As Salazar Mostajo (1989: 80) puts it, the murals depicted 'an Indian that corresponded more to the image of the hero of the siege of La Paz than to the *pongo* who served in the hallways'. The representations 'did away with all meekness, all delicacy; the Indian is shown as forceful, ready to explode, as given a power to enter the action...' (ibid., author's translation).

Figure 4.1   Tempera mural at the Warisata School. Alejandro Mario Yllanes.
Photograph Courtesy of Laura Salazar de la Torre

Figure 4.2   Tempera mural at the Warisata School. Alejandro Mario Yllanes.
Photograph Courtesy of Laura Salazar de la Torre

In a way, the Warisata project proved to be a victim of its own success. 1934 saw a wave of uprisings by landless peasants or *colonos* which 'spread from La Paz to Oruro, Potosí and Sucre' (Hylton and Thomson 2007: 69) and rumours of the school's influence as a font of activism was more than sufficient to turn it into a target of reprisal from landlords and political elites.

> Just as the school's fame was spreading beyond Bolivia to Peru, Mexico, and the United States (which sent streams of visitors to observe this innovative school)... Bolivia's conservative aristocracy accused Warisata's teachers and *amawt'as* of preaching racial hatred, walling off the Indian school from national society, and stirring up the restive peasantry across the altiplano. But the school also came under attack from a faction of progressive educators, who denigrated the school's underlying philosophy of cultural pluralism and instead pushed forward an assimilationist agenda, aimed at converting Aymara (and Quechua) Indians into generic (culturally 'mestizo') *campesinos*.
>
> (Larson 2011: 65)

As Hylton and Thomson (2007: 70) note, in the late 1930s and early 40s, 'the Indian question' was eclipsed by a broader wave of protagonism from 'increasingly radicalised miners, rural tenants, students and the nationalist parties and regimes tied to them'. The loss of the Chaco to Paraguay and the staggering numbers of dead and injured led to much recrimination at home. Bolivian leaders were seen to have 'precipitated another extensive and humiliating dismemberment of their country' (Morales 2003: 108) and many of those who had survived the war wanted to change the country. A crisis of political legitimacy unfolded. In this context, the traditional oligarchic parties were quickly upstaged by the emergence of a new mass politics and a range of new parties emerged, all unified in their aim to break down the established order. Of these, four were to play an especially influential role in post-war politics: the *Partido Obrero Revolucionario* (Revolutionary Workers Party or POR), established in December 1935; the *Falange Socialista Boliviana* (Bolivian Socialist Falange or FSB) established in 1937; and later, the *Partido de Izquierda Revolucionaria* (Party of the Revolutionary Left or PIR) set up in 1940 and *Movimiento Nacionalista Revolucionario* (National Revolutionary Movement or MNR) which emerged in 1941.

Scheina (2003) writes that one of the casualties of the Chaco War was Bolivia's fledgling democracy. The years following the war saw a series of *coup d'etats*. In 1939, Germán Busch Becerra proclaimed himself dictator. Through the Office of Indigenous Education, Busch had promoted the Warisata project, envisioning it as part of his 'uniquely Bolivian' national socialist project. However, following the dictator's suicide in 1939, the oligarchy once again seized control of national government and in 1940, Bolivian public education was revamped and extended as the state's arm of Indian assimilation. In this context, Warisata was shut down and transformed. Its *tempera* murals survive only in the documents and photographs of Carlos Salazar Mostajo (reproduced here).

The National Revolution and 'the social painters'

The next 12 years of Bolivian history were extremely turbulent, featuring coups, assassinations and the formation of various unholy alliances. Over this period, the voting public progressively expanded and became more polarised; the ideological foundation of Bolivian politics also shifted, making way for the MNR to emerge as a multiclass party of the left. Morales (2003: 133) writes that the MNR found its political identity in the slogan 'land to the Indians, mines to the state'. On the one hand, despite the best efforts of Indian activists and criollo reformers, 'Bolivian society and its power structure were [still] cast in the rigid mould of the Spanish colonial institutions' (Patch 1961: 124). Up to 95 per cent of land was still held in large estates of over 10k hectares, owned by descendants of the Spanish elite. On the other hand, the low price of tin (arguably set to beseech to the US government) and government repression of strikes[5] drew impoverished miners into the opposition movement led by the MNR. Patch (1961) reflects that by 1952, tensions were so high that if Bolivia did not have an MNR government, it would have had no government at all. Dubbed 'reluctant revolutionaries', the MNR swept into power by force in 1952 after having been denied its legitimate victory in the 1951 presidential elections (Hylton and Thomson 2007).

Having returned from exile the new President, Victor Paz Estenssoro, instituted a series of transformative reforms. He immediately introduced universal suffrage, removing requirements based on literacy and affluence. In so doing he expanded the electorate from around 200,000 to more than one million, newly enfranchising many Indians and women but he also attempted to reorganise society along class lines. In essence, the new MNR government provided concrete incentives to turn *ayllus* into agrarian unions or *sindicatos* that would institutionalise support for the party (Heath 1955, cited by Yashar 2005). It established the Ministry of Peasant Affairs and an agrarian reform programme inspired by the Mexican model was implemented, enabling unionised Indians – now regarded as *campesinos* (peasants) – to reclaim tracts of land from large latifundio-type holdings (Yashar 2005). Paz also nationalised the mines, combining the mines owned by the three wealthiest tin barons under the umbrella of *Corporación Minera de Bolivia* (COMIBOL), the new state mining company. The miners' movement maintained a strong level of influence in government, co-governing mine administration through the *Central Obrera Boliviana* (COB).

Notably, Paz and other key exponents of the National Revolution understood and promoted public art as a tool of national political transformation (Salazar Mostajo 1989). The new government set aside financial resources for increased cultural activity and was especially encouraging of a burgeoning muralist movement which came to be known as 'the social painters'. The group included such figures as: Walter Solón Romero; Miguel Alandia Pantoja; Lorgio Vaca and Gil Imana. In 1953, Victor Paz Estenssoro invited the Mexican muralist Diego Rivera to Bolivia. This visit had a strong influence on the direction of public art and initiated a rich exchange of ideas between the two countries about the role of the aesthetic in both forging a new cultural identity and nurturing the new mass politics (Arandia 2013a). The 'social painters' drew on methods, themes and styles trialled in Mexico and beyond by the *tres grandes*. Miguel Alandia Pantoja, a self-taught artist from Catavi, found great inspiration in the work and style of José Clemente Orozco. He had served during the Chaco War and had subsequently been imprisoned in Paraguay where he came to understand the war as a despicable endeavour hatched by vying elites at the expense of the poorest members of society. Upon returning to Bolivia, Alandia became an active militant Trotskyist with the *Partido Obrero Revolucionario* (POR). During the *sexenio* – the six years prior to the revolution – he worked with others to found the National Workers Union, the immediate predecessor of the COB (Montoya 2007). In 1952, he marched on the capital, rifle in hand, to demand a better standard of living for the workers and the dispossessed masses.

Although his politics diverged from the nationalists of the MNR, Alandia's militancy guaranteed him a platform and space in the cultural programme of the revolutionary government (de Oliveira Andrade 2006). Alandia's murals celebrated the workers, portraying their importance to the economy by rendering them on a large scale. In Alandia's words, his work pays homage to the 'myths and legends, the lives of miners and peasant masses in their struggle against the old landed and commercial mining oligarchy, [it seeks] to express in a visual language, the rejuvenated and resounding universal longing of man in our times: revolution' (Alandia, cited by Montoya 2007). In 1953 Alandia painted the mural '*Historia de la Mina*' (History of the Mine) in the Governmental Palace. The mural stood at an enormous 86 square metres in size. His other murals included: '*Historia de la Medicina*' (History of Medicine), a mural of 50 square metres in size, which was made on canvas initially and then installed in the Worker's Hospital of La Paz in 1956; a series of five murals painted on the building of the state company YPFB, which took '*El Petroleo*' (Bolivian oil) as their subject and the 1964 production entitled '*Lucha del Pueblo por su Liberación, Reforma Educativa y Voto Universal*' (The People's Struggle for their Liberation, Educational Reform and Universal Suffrage), an extremely large-scale mural commemorating the 1952 revolution which can be seen from the Plaza Villarroel in La Paz.

Yet, Alandia viewed painting as just one part of a larger project of political agitation and critique. As a result, de Oliveira Andrade (2006) writes that it was hard to separate the man as activist from the man as artist. In addition to systematically seeking to disseminate his visual messages of revolution in cultural centres and trade unions, Alandia collaborated intensely with party newspapers and trade union organizations, contributing a variety of drawings and prints centred on the activities of the working class. Alandia's work was particularly well received by Diego Rivera who was shown the '*Historia de la Mina*' in 1953. Following this trip to Bolivia, Rivera proudly exclaimed: 'there is now a movement of monumental collectivist art across our continent!'; '[Miguel Alandia Pantoja's] production is a clear indication that our movement has become the instrument of artists who produce their work side-by-side with their countrymen' (Tibol 1957).

Walter Solón Romero meanwhile, was a formally trained artist who eventually became well known across Bolivia for his murals, frescos, and engravings. From

Figure 4.3   'Historia del Petroleo Boliviano'. Walter Solon Romero. 1958.
Image courtesy of Fundación Solón

early on in his career, Solón had committed to a form of social art that aimed to denounce the injustices afflicting people in their day-to-day lives. Following the Revolution, however, he was commissioned to produce a series of monumental works of allegory, beginning with the '*Historia del Petroleo Boliviano*' (History of Bolivian Petroleum), which was painted in the nationalised Petrol Industry offices in 1958. Montoya (2007) writes that the Mexican muralist David Siqueiros had been a strong influence on Solón. The pair had met during the 1950s, in Chile and, '[l]ike those of Siqueiros, Solón's murals are meant to reach out and communicate with his people, to capture its history and its sensibility' (Sacks da Silva 2004: 96). Hence, the *Historia del Petroleo Boliviano* juxtaposes different phases in the history of Bolivian oil extractivism: the discovery of oil in colonial times; concessions made to the US company Standard Oil; the national mobilization to defend of Bolivia's natural resources during the 1932–1935 Chaco War; the founding of YPFB on 21 December 1936; and, the first nationalization of hydrocarbons in March 1937. It also includes a vision for the future in which better living conditions prevail for the Bolivian people.

As with the Mexican muralists, Bolivia's 'social painters' crafted epic murals on the walls of highly visible public buildings. They developed an iconography featuring heroes from the nation's past, its present, and its imagined future. This included Indian warriors resisting colonial rule, peasants and miners coming together to march in 1952, as well as labourers busily working to forge an imagined great state of the future. Like the work of Yllanes 20 years prior, the works of Solón, Pantoja and the other muralists of the revolutionary era took on a didactic and instrumental function, seeking to advance a project of political transformation and incorporate a new political class by raising civic consciousness and making the 'everyday man' a celebrated figure in the history of the nation. However, unlike the earlier, more pastoral, muralism of Yllanes, the 'social artists' worked with images of a past, a present and a future in which industrialism and extractivism play a prominent role. In revolutionary muralism, the Indian is everywhere: appearing as soldier, miner, pilot. S/he is a part of modern society. But, less common are the self-sufficient farmers, fishermen and tanners of Warisata and the semi-nomadic indigenous people of the Bolivian Amazon.

Patch (1961: 124) offers some explanation for the shift in focus. He claims, '[t]he revolution not only placed the land in the hands of the men who cultivated it, it also destroyed the institution of the *latifundio*, and went far towards replacing the castelike status of the Indians with a class concept of the campesino [peasant] in which mobility is possible'. Hence, the social mobility of the Indian becomes an achievement to be celebrated in visual form. However, the attempt to organise the indigenous population from above on a non-ethnic basis – to identify them as peasants and treat them as a social class (Hammond 2011: 650) was problematic in certain respects. In many instances, the new arrangement only inserted union structures into pre-existing indigenous community authority systems. Hence, the *sindicatos* 'failed to eliminate traditional indigenous organisations such as the ayllu from the institutional landscape of Bolivia. Pockets of traditional leadership and organisational forms persisted in regions such as Oruro, Potosi, Chiquisaca, Beni and Santa Cruz, despite persuasive political, social and economic pressure to weaken and marginalise them. In some areas … the sindicatos had served outward relations and representation, while the ayllu handled the "internal" matters of indigenous and community affairs' (Healy 1996, cited by Yashar 2005: 162). Moreover, by incorporating the *campesino* as part of the class system, the reforms relegated and marginalised the lifestyles of those Indian communities existing largely outside of the capitalist economy, fomenting a division that would come to shape the new politics of *indigenismo* in the twenty-first century.

> Not only did the agrarian reform law disregard the interests of Amazonian Indians, it actually referred to them as wards of the state rather than as capable and full-fledged citizens. Article 20 of the 1953 Land Reform, for example, declared that '[f]orest groups of the tropical and subtropical plain that find themselves in a savage state and that have a primitive form of organisation, will remain under the protection of the State.
>
> (Yashar 2005: 194)

However, it would not be fair to say that this was the opinion of the social painters, nor would it be accurate to say that work of these muralists was merely an extension of the state's aims and ideology. There were strong disagreements between the individual artists and there were also tensions between the artists and the state. Initially Pantoja deplored Solón, who was a supporter of the MNR and regarded him as a fair-weather activist and lackey of the bourgeoisie (de Oliveira Andrade 2006). The two artists clashed politically and this was reflected in battles over prominent public spaces in which each desired to paint. A dispute over space at the *Museo de la Revolucion* was only resolved by allowing the two artists to share the wall in equal parts. 'The controversy surrounding this and other murals painted … in public buildings and unions was intense, and it reflected much deeper social tensions over control of the historical record of the revolution through art' (ibid.).

There is also rather more to the murals than meets the eye. Walter Solón's son Pablo recounts that his father kept many secrets in his murals. He used them to portray his wife, his grandchildren, his children and his friends. Indeed, his entire family would pose for him and be written into his allegories. Solón would also use his painting to play pranks: in one work he painted his close friend Pepe Ballon who was at the time, a little old man. However, Ballon is portrayed as a small child in the midst of the 1952 revolution. Pablo states: '[The picture] was inspired by the social struggle, but at the same time it was a product of tenderness and joy. It takes the opportunity to make mischief out of life.' Moreover, Fundación Solón (2011) describe several recurrent motifs in Solón's art. In some works a stone or rock with eyes can be deciphered. Solón's rock with eyes was intended to remind Bolivians of the importance of history and memory. A winged figure named *Anteo* also features repeatedly. Anteo's feet always remain firmly rooted on the earth and this was Solón's way of saying that realism and humility were essential to all political and social visions (ibid.).

The most frequently recurring figure or icon in Solón's murals is Don Quixote, a character from the famous works of Cervantes who, according to the novels, is an older gentleman that becomes obsessed with the chivalrous ideals of fairytale knights and then decides to take up a sword and steed and go off in defence of the helpless and oppressed. The stories of Quixote are well known across the Americas, making him an immediately recognisable figure in the murals, symbolising 'the tireless fighter for justice in a world that considers him mad' (ibid.). However, Solón's fixation with the figure of Quixote had a personal, nostalgic and affective dimension too:

> All of you have to wonder why the Quixote is among us? Why is it that he intrudes in our reality and is almost always troubled to the extreme, coming to identify with an Indian or mestizo?
>
> This is the story… A vision emerges from memory: a village, my house, my brothers, a big table, a canvas spread like salt on the pampas, and the silhouette of a man who is distracted or distracting us making drawings on rolls white paper. It was my father who at night drew great figures with coal black as the night. Outside of the house, the wind became the musical background to the faces, scenes and landscapes that arose [on the paper]. One night my father drew a very thin man dressed in armour and holding a spear. He said it was Don Quixote. He added a squire, Sancho Panza and after that a steed, Rocinante, who was as thin as the man with the sad figure. Simultaneously [my father] told us [Quixote's] adventures and misadventures on a never-ending basis. Since then, when I was very young still, I used to associate the lean and slim figure of a man with strength, goodness and justice. I had not learned to read and yet I knew Don Quixote…
>
> (Solón n.d.)

While it is rather easy to dismiss revolutionary muralism(s) offhand by claiming that such overtly committed art forms do little more than promote a particular ideological slant (see Gao Xingjian, cited by Bleiker 2009), the recurrent figure of Don Quixote in the murals of Solón, for example, can be pressed to reveal so much more. For Solón, Quixote is perhaps as much of an autobiographical stamp and marker of personal memories and sentimental attachments as he is a public and political symbol. Untangling the personal and the political here is a challenging, if not futile, endeavour.

Contra la dictadura: street art as resistance to authoritarianism

After deepening Bolivia's dependence on US tin contracts and development aid (both conditional on a commitment to anti-communist counterinsurgency) over the 1950s, the civilian government of Paz Estenssoro was toppled in 1964. After the coup, René Barrientos dismissed claims that he would roll back reforms instituted as part of the revolution. In fact, he emphasised continuity, stating that he would 'restore the national revolution' (Duane Lehman 1999) by bringing increased prosperity and autonomy to Bolivia. In practice, Barrientos was pro-market, anti-union, staunchly anti-communist and backed by the United States. However, he did secure a strong support base among the new *campesino* class by travelling to rural areas and addressing the people in Quechua. He engaged in a clientelist model of politics dubbed the Military–Peasant Pact, building local schools and medical outposts in return for political backing. In 1967, Barrientos guaranteed his place in international history by presiding over the capture and assassination of the revolutionary Ernesto 'Che' Guevara and his column in Vallegrande. One of the oft-cited causes for the failure of Guevara's campaign was the lack of buy-in from the Bolivian peasantry who were neither as destitute or neglected as their Cuban counterparts (Johnson 2006).

Rather unsurprisingly, the military government of Barrientos and those that followed in his wake were not enthusiasts for the political and revolutionary message of the murals painted by Alandia Pantoja, Solón and the other social painters. Many of these works, which immortalised the actions of the Bolivian people in the revolutionary struggles of 1952, were destroyed by the military governments that followed after 1964. Alandia Pantoja's '*History of the Mine*' (1953), and '*History of Bolivian Parliament*' (1961) were both destroyed in May 1965. de Oliveira Andrade (2006) writes that the close relationship that had been forged between the muralists and the labour movement was clearly demonstrated in the Federation of Miners' response to the impending destruction of Alandia's works:

> It is the revolutionary duty to defend the work of art, above all ideological or aesthetic consideration. It is inconceivable that the Alandia murals are covered with white paint … If we get to the ungrateful extreme [of] deciding the destruction of Alandia murals, the FSTMB is willing to [move] them to its headquarters before allowing such an act of vandalism.
>
> (Lora, cited by de Oliveira Andrade 2006)

Following the mysterious death of Barrientos in 1969 and a brief socialist interlude under the ill-fated Juan José Torres,[6] General Hugo Banzer Suárez seized power in August 1971. Where Torres' brief intermission had created political space for indigenous organising, Banzer antagonised emerging groups like the *Kataristas* who had begun to forge a new collective identity based on the complex reality of intersecting ethnic and class exclusions in post-revolutionary Bolivia (Yashar 2005). Banzer initially had the support of two of the country's leading political parties but assumed unlimited powers in 1973 when he replaced civilian politicians with military personnel. Banzer's regime ruled without interruption until 1978. In this time it banned forms of political organisation, closed down universities, dispensed with freedoms of expression and forced all major opponents into exile, including many young *Kataristas* (Morales and Sachs 1987). Banzer also cooperated with other military generals from the region – Chile's Augusto Pinochet, Alfredo Stroessner of Paraguay and the Argentine General Jorge Videla – in sharing intelligence and secretly transferring prisoners under the auspices of *Operación Cóndor.* In effect, the 'Banzerato' was a period of extreme repression, violence and fear. In January 1974, for example, up to 200 Indian peasants from Cochabamba were killed when police opened fire on them for protesting at the removal of concessions granted in the 1952 Revolution. Yashar (2005) shows that during Banzer's reign, public protest events dropped significantly in number as collective action became increasingly risky.

Aymara artist and curator Edgar Arandia Quiroga (2011) relays that during the Banzer period the topic and image of both the Indian and the revolutionary were erased from the art world and the mainstream media almost entirely as the government dictated rigid guidelines to the country's cultural and media institutions: the Indian should not be encouraged to rebel. Banzer ordered police to destroy the remaining murals of the social painters, as a show of zero tolerance towards 'revolutionary spirit'. The police removed paintings by Diego Rivera from the museums for fear that they would stir popular opposition. Muralism was, in effect, outlawed.

However, the hard line of the government did not entirely eliminate forms of activism. In the Bolivian Amazon, where collective organisation was harder for the state to track and clamp down on, indigenous groups including the Izoceños-Guaranies, Ayoreos, Guarayus and Chiquitanos came together to form the *Confederación de Pueblos Indígenas de Bolivia* (Confederation of Indigenous Peoples of Bolivia or CIDOB) and began a process of slowly unifying lowland groups. Meanwhile, in urban centres such as La Paz and Cochabamba, artists were taking their more politicised works underground. Walter Solón, for example, continued to produce his *Quixotes* using materials and mountings that could be moved or copied and distributed with greater ease. He claimed that his characters had become witnesses and commentators on the violence of the regime and it became imperative to preserve them as a record of the times. Solón's drawing '*Quixote y los perros*' (Quixote and the Dogs), for example, was intended as a denouncement of state violence. It was disseminated widely after the disappearance of his eldest son at the hands of the regime (Fundación Solón 2011). Others, including the young Edgar Arandia, decided to take to the streets during this period, using more ephemeral forms to demonstrate the persistence and intractability of the resistance.

Clandestine art interventions were sometimes made in daylight but with much haste. Arandia (2011) explains that these actions would sometimes be the initiatives of smaller cliques within a group named the *Círculo 70*. The *Círculo 70* was founded in 1972, by a group of young art-activists who had become angered by repression and censorship under the regime and felt that Hugo Banzer's close relationship with the US government was detrimental to national autonomy and social justice. The *Círculo* consisted at one time or another of the artist-activists: Silvia Peñaloza; Edgar Arandia; Mario Velasco; Juana Encinas; Inés Nuñez; Erasmo Zarzuela; Ricardo Pérez Alcalá; Benedicto Aiza; Windsor Arandia; Héctor Terceros; Emilio Tórrez; Enrique Pacheco; Froilán Argandoña; Aydé Aguilar; Cristina Endara; Gíldaro Antezana; Julio César Téllez and Luis Ángel Aranda (El Diario Cultural 2012). *Círculo 70* meetings often occurred in secret. At these meetings, students, academics, artists, former unionists and other dissenters would come together to strategise acts of insubordination. They designed and disseminated fliers and pamphlets – what Chaffee calls 'auxiliary modes of street art' – containing illustrations and poetry intended to remind people that they should not be obliged to live in fear (Arandia 2011). They also devised slogans and symbols which were disseminated across urban centres by means of flyposting, pamphleting and graffiti. Some slogans would critique the regime for its perceived subservience to the United States: '*Patria o muerte*' (Homeland or death); '*Muerte US imperialismo*' (Death to US imperialism). Some recalled injustices and abuses of power: '*Libertad para los prisoneros*' (Freedom for the prisoners). Other common refrains called for an end to the dictatorship: '*Muerte a la dictadura*' (Death to the dictatorship); '*Viva la revolución popular*' (Long live the popular revolution); '*Viva la democracia*' (Long live democracy).

Arandia (2011) recalls that graffitied slogans were frequently painted over by the authorities, who tried hard to maintain the imaginary of the street as a tightly ordered space. The anti-regime slogans kept the idea of political opposition very much alive and present in peoples' minds, even if there was no possibility of expressing that opposition at the ballot box. Moreover, the visual noise generated by the graffiti undermined the regime's attempt to project a public image of dominance and order. Reflecting on political graffiti and its role in the politics of resistance, Arandia (2011) suggests that there was satisfaction to be gained from watching the authorities engaging in the repetitive and mundane task of whitewashing the city walls. Perhaps for this reason slogan-painting became an almost self-sustaining practice in La Paz and El Alto: 'as soon as inscriptions were covered over in one location, several more would pop up elsewhere. It was impossible to subdue them.' State and activists remained locked in this dance, with neither side surrendering.

Although they remained illegal, political murals also began to reappear in the late 1970s. Now rather more of a bottom-up enterprise, political murals would emerge in abandoned stations and alleyways – nooks and alcoves where the risk of being caught painting was lessened. Around the ten-year anniversary of the death of Che Guevara, a number of large-scale colour murals of the guerrilla fighter appeared across spaces in El Alto. Although Guevara's Bolivian insurgency had failed to mobilise significant support at the time and had ultimately ended in his death, through a combination of 'myth and memory' (Hylton and Thomson

2007: 85), his *Ejército de Liberación Nacional* (War for National Liberation or ELN) later became an inspiration for activist groups, including the student movement. Guevara emerged as a symbol of popular defiance to capitalism, US imperialism and perhaps most importantly, the military institution that had earlier orchestrated his killing. In Vallegrande, where Guevara was assassinated, the image of Che gradually emerged as part of church iconography, appearing alongside Christ and the saints (Kunzle 2006).

Other street-based expressions had greater uptake among activists during the Banzerato and after, as García-Meza came to power. One of these was street theatre. In La Paz, dissidents were often resource poor, lacking both financial options and access to channels of communication. The virtue of street performances was that they required nothing more than one's own body and a bit of creative impulse: a crowd could quickly assemble in a plaza or street to watch improvised political satire or re-constructions of military crimes, and just as quickly, an audience and its performers could disperse and melt into the city (Arandia 2011). Additionally, underground magazines, pamphlets and other ephemera were produced in the early 80s including the publications, *Trasluz, Comprada Mouser* and *Papel Hygenico*. These highly critical publications were dispersed on the streets and during meetings of resisters. In the absence of an open and pluralistic political system, Chaffee (1993) argues that the production of underground ephemera serves an important role in breaking the complicity of silence under authoritarianism. Arandia's reflections support this view. He asserts that, 'the distribution of magazines and pamphlets were important. There was no freedom for journalists, no critical voice. These publications gave an opposing view, making a mockery of the military, the president and his circle of allies' (ibid.). However, there is increased risk associated with the setting up of underground publishing houses as opposed to the more improvised and/or low technology modes of expression. Magazine publications require a more or less permanent physical space for equipment to be used and stored as well as financial backing that can often be traced back to particular individuals or groups. There is also a literacy requirement placed on the readers, which can, in certain instances be exclusionary. According to the UNESCO Institute for Science and Statistics, Bolivia exhibited an adult literacy rate of 63.2 per cent in 1976, growing to an estimated 66 per cent by 1981 (IndexMundi 2012). One virtue of aural and visual expressions on the street was their accessibility at all socio-economic and literacy levels.

Arguably, the impulsion and fervour with which authoritarian regimes seek to suppress oppositional street art and immobilise those who create it are testament to the very power of the medium. Under Banzer and later García Meza, punishment for producing dissenting street art and its auxiliary forms ranged from jail to torture or even death. The director of *Trasluz*, Rene Bascopé was assassinated. Walter Solón was detained and tortured by military officials who threatened to cut off his hands. Members of the *Círculo 70* and other dissident groups also suffered for their creative endeavours. Diego Morales was captured, tortured and eventually forced into exile (Inter-American Commission on Human Rights 1982). Edgar Arandia meanwhile was shot and then went to exile, travelling to Ecuador, Haiti,

Mexico, Germany and Spain before returning to La Paz in the democratic era to take up a position as an art director:

> The punishments for actions against the regime were severe. I became known to the government for my activities in the art circuit and on the streets. They shot me in the stomach and had to leave the country for fear that the next time I came into confrontation with the regime, I would be dead.
>
> (Arandia 2011)

When asked why, in the face of such violent retribution, he continued to create subversive art, Arandia (2013b) reflects: *'First: passion, excessive passion. Because if you begin to consider up to what point you are going to be a painter or up to what point you are going to be a human being, you are lost...'*

The 'death of politics' and the rise of anarcha-feminism

After a chaotic period of coups and counter-coups, civilian rule was finally restored to Bolivia in 1982, with the collapse of the military government and the accession to power of Hernan Siles Zuazo, who had years earlier broken ties with Victor Paz and the MNR to form a new party, the *Unidad Democratica y Popular* (UDP). With the political opposition legalised once more, political campaigns regained considerable visibility on the streets. Printed posters and painted slogans instructing the populace to vote for this or that candidate, party or outcome bedecked concrete walls and shop windows. Politicians and their representatives canvassed energetically. Hylton and Thomson (2007: 90) write that Siles Zuazo 'took charge amidst high expectations and a robust sense of popular power' which saw considerable expression through street art. However, this optimism was short-lived. Zuazo inherited an economy already on an inflationary path, with price rises reaching hyperinflationary proportions in April 1984. The institution of exchange rate and price controls, increased export taxes and other measures intended to curb inflation, had high welfare costs, leading to rapid expansion of the informal economy. The changes were met with widespread strikes organised by the unions. UDP attempts to negotiate with and appease business groups inspired uprisings from other sectors of society. Hylton and Thomson (2007: 90) note that, 'The UDP government emitted ever more currency to cover the cost of the agreements it signed with unions and civic committees, and in the early 1980's Bolivia became the country with the highest inflation rate in the world'. In May 1984 the government stopped servicing foreign debt and in the face of crisis called for elections.

Víctor Paz returned to the presidency in 1985 after fighting an election against former dictator, Hugo Banzer. Paz was considered the more moderate candidate so it was a big surprise to the public when he sought to implement Banzer's neoliberal 'shock' programme, the New Economic Policy (NEP). The NEP involved removing exchange rate controls and barriers to trade in order to attract FDI back to the country, renegotiating the external debt with international creditors, denationalising key sectors and cutting state payrolls. The government devalued the peso from 75,000 to the dollar to one million to the dollar, making imported goods completely unaffordable for much of the populace. Problems were compounded by the fall in tin prices on the international markets. Tin, which had provided a great proportion of government revenue since before the 1952 revolution, was suddenly turned from Bolivia's most lucrative product to its biggest liability. Mines were privatised as a coping measure and more than 22,000 miners from COMIBOL – most from indigenous communities – were 'relocated' from Oruro and Potosí under Supreme Decree 21060 (Hylton and Thomson 2007). Many of this number were forced to fend for themselves in the informal economy. Others migrated to the Chapare region and began to produce *coca*, a sacred leaf used for ritual and daily consumption in the Andes.

Many indigenous ex-miners eventually became 'politically active' in the coca growing regions of the Chapare (this is hardly surprising given the labour radicalism of the miners movement). However, in the intervening period, Paz Estenssoro's measures had the effect of breaking the 'backbone of the country's radical workers' unions' (Dangl 2010). Hylton and Thomson (2007: 96) elaborate:

> Arguably Latin America's most combative proletariat in the second half of the twentieth century, the tin miners were broken when Paz Estenssoro – who had first risen to power on the strength of the miners' militias in 1952 – crushed their 'March in Defense of Life' in 1986. The FTSMB and the COB would never again have the same capacity to agglutinate a broad array of forces around a national popular program.

Left political parties entered decline as initial hopes of a transition from 'dictatorship' to 'democracy' to 'socialism' faded. As María Galindo (2010: 55) describes it, '[t]he miners' movement collapsed like a house of cards, and that brought down the entire popular movement. It practically signaled the death of politics, the politics of resistance.' While crippling the poor and disempowering civil society and the unions, the NEP was viewed as a resounding success by international observers from the IFI's. There was little incentive for the domestic government to change tack. It was at this time, when the social movements of Latin America were still sleeping and neoliberalism was in full bloom (Dangl 2010), that the anarcha-feminist group *Mujeres Creando* emerged.

Mujeres Creando (Women Creating) has its origins around 1985 when three activists Julieta Paredes, María Galindo and Mónica Mendoza decided to initiate 'a "revolution" in the use of public spaces such as the street' (Galindo 2010: 55). The three began their political lives more or less aligned with the parties and movements of the Bolivian left. In concert with their contemporaries, they were dismayed by the choice of economic policies undertaken by the civilian government and they were angered by the continued impunity for characters like Banzer, who had been reabsorbed into the political system in spite of abuses committed during his period of military rule.

At the beginning the group focused on the legacy of the dictatorship and then, the widening chasm between the people and their elected government: '[u]p at the top, in terms of public order, the politicians were deciding things, but the people were actually sorting out their stuff on the streets' (Galindo 2010). In effect, the demobilisation of traditional institutions such as the trade unions and the political parties, coupled with rising unemployment and informalisation effectively made the streets the most important site of struggle. This struggle did not involve a scramble for political power but rather aimed to carve out new modes of continuation, subsistence and progress amid rising levels of socio-economic strain. A number of scholars working on the effects of neoliberalism in Latin America (Robinson 2009; Hite and Viterna 2005) have registered parallel examples of 'non-hierarchical grassroots organising … largely independent of political parties' that treated the street as 'the new factory' (Robinson 2009: 297). In Bolivia, Galindo, Paredes and Mendoza observed how in many ways, it had been women that led the charge to take back the streets, strengthening the social fabric by co-operating and organising to ensure that material needs were being met. In this way, women were essentially converting public spaces into domestic spaces, building on some long-established traditions of making and doing that were consistently overlooked by the patriarchal political institutions (Paredes 2002).

One need look no further than the everyday lifestyles and practices of *cholas* – or *cholitas* as they are more affectionately known – for evidence of the mobilising strength, political imagination and pragmatism of women in Bolivian culture and society. The word 'chola' comes from the Spanish pejorative for Indians of mixed descent. In Bolivia, Aymara and Quechua women have appropriated the term – and the European clothing that the Spanish forced them to wear[7] – 'as a badge of honour' (Dear 2014). '[I]ndependent of their husbands and of the wage-labour market' (Seligmann 1989: 705), Bolivia's *cholitas* subsist on the surplus they can generate through moving and trading goods between the rural communities and urbanites engaged in the formal capitalist economy. Historical sources place *cholitas* at the site of many political protests, where they have often rallied in support of peasants and rural workers. For example, after General Luis García Meza's coup of 17 July 1980, *campesinos*, miners and *cholitas*, participated in extensive and disruptive uprisings across the country. They 'built blockades to prevent the flow of goods to and from rural areas' (Seligmann 1989: 715). These actions had the desired effect of (at least temporarily) weakening the power base of the new military junta. In the 1980s, rather than accelerate the assimilation of *chola* women, neoliberal reforms boosted their role and importance. *Cholitas* became an essential provider of goods and services for a much expanded and increasingly impoverished urban and informal workforce (Seligmann 1989).

Rather than the *death* of politics then, Galindo, Mendoza and Paredes observed a relocation of politics as a result of neoliberal programming: 'the street became the most important survival space, the most important forum for the whole of society' (Galindo 2010: 55) and women were indispensible to this. They argued that if the Bolivian left wanted to regain traction and relevance in the post-transition context, it would need to return to the street and actually engage with women. Yet, Paredes, Galindo and Mendoza struggled to get this point across in meetings and demonstrations. It soon occurred to them that while women provided essential numbers and votes in the party meetings, they were thought of more as window-dressing than as equal partners in building a movement for change. Accusing the organised left of arrogance, homophobia and totalitarianism, the women formed a splinter movement of their own (Paredes 2002).

The founders of *Mujeres Creando* describe the collective as 'a craziness' dedicated to challenging the injustices in Bolivian society. These ranged from economic injustice and impunity for political violence, through to racial and gender-based discriminations. The collective includes some of Bolivia's few openly lesbian public figures, anarchists and feminists:

> [We are] crazy people, agitators, rebels, disobedients, subversives, witches, street, grafiteras, anarchists, feministas, Lesbians and heterosexuals; married and unmarried; students and clerks; Indians, chotas, cholas, birlochas, and señoritas; old and young; white and coloured, we are a fabric of solidarities; of identities, of commitments, we are women, WOMEN CREATING.
>
> (Anon. cited by Ainger 2003: 256)

Determined to address the growing gulf between the Bolivian people and existing political institutions, the collective set out to animate the streets, interject in the everyday lives of citizens and foster new feedback mechanisms. Right from the outset, they used graffiti to denounce forms of oppression and 'excommunication', developing a distinctive, cursive tag, vaguely reminiscent of the banners and posters of the *Madres of the Plaza de Mayo*. As one member of the *mujeres* explains, the group's impulse toward these street-based *pintadas* (paintings) transpired from a desire to subvert and refuse the one-way channel of communication between political or commercial elites and the broader populace (Paredes 2002).

To take an example, in the run-up to the 1985 elections candidates distributed instructional handbills across the city of La Paz. Rather than offering voters reasons for choosing this or that candidate – information drawn from their manifestos, for instance – the handbills simply replicated ballot papers. They would display a list of candidates' names, with one of these highlighted for posterity. The University of New Mexico hosts an online catalogue of handbills from the period which all follow this pattern. One sample from Jaime Paz Zamora's campaign contains his portrait and logo, the rooster. It issues a simple instruction to the voter: *'busca el gallo y marca en el cuadrado blanco'* (look for the rooster and mark in the white square). For the *mujeres*, these instructional campaign tools did not reflect the values of an emergent democratic system at all. Graffiti became a 'response to [parties] painting in the streets, saying "vote for so-and-so". [These] were affirmative or negative phrases: "No to the vote", "Yes to this", "No to that"' (ibid.). In other words, the collective used their paint to 'talk back' to political elites; to highlight the discriminatory consequences of proposed or existing policy, illuminate political corruption and call out authoritarian and patriarchal elements persisting within new the civilian machinery. Slogans challenging the subordinate status of women in society have included: *'La*

Figure 4.4 'Putita' from Slutwalk organised by Mujeres Creando, 2011.
Author's own photograph

mujer es la proletaria del proletario' (The woman is the proletariat of the proletariat); *'Mujer que se organiza plancha más camisas'* (Women that are organised iron more shirts); *'En árabe aymara y castelló, mujer quiere decir dignidad'* (In Arabic, Aymara and Spanish, woman ought to mean dignity'); *'Confío en el sonido de tu propia voz'* (Trust in your own voice); *'El Mercado es mi casa sin marido'* (The market is my home without a husband). Meanwhile, phrases such as: *Civismo rimas con fascismo y racismo* (Civic spirit rhymes with fascism and racism); *'Ser maricón es una opción, ser corrupto es una degeneración'* (To be gay is an option, to be corrupt is a degeneration); and, *'No hables de revolución con Goni, solo un vacación de Justicia'* (There's no talk of revolution with Goni, just a vacation from justice)[8] challenged other deficits and weaknesses in the democratic system.

As McLuhan and Fiore (1967: 69) have put it, 'environments are not passive wrappings but are rather active processes which are invisible'. They continue, 'the ground rules, pervasive structure and overall patterns of environments elude easy perception' but they exert a strong influence over the way we act. In McLuhan's view especially, artists have a particular skill for initiating 'anti-environments' which call these structures of power into focus. In other words, through art interventions, it is possible to raise public attention to the unconscious or unchallenged exclusions that characterise a given environment at a given moment. Alongside the top-down communicative practices cultivated by political elites, the *mujeres* also cited the creeping privatization of public spaces as an active, invisible and anti-democratic process: 'if we must pay for public space, then it's a big contradiction in democracy' (Paredes 2002).

The group's unsanctioned *pintadas*, in effect, reclaimed physical space from the commercial and political elites. In so doing, they rearticulated the boundaries for political practice and challenged the prevailing urban aesthetic. In the following statement, Paredes expands on this, describing 'the aesthetic' not as an objective and abstract domain of critique but rather as a product of standpoint, positionality and power. She posits, 'Oh, so when Coca-Cola contracts a painter, it doesn't make the wall dirty? ... It seems to me that it has made the wall dirty in a disgusting way. And what we have done, our graffiti, that's beautiful' (ibid.).

Paredes' comment resonates to some degree with Bourdieu's work on taste and habitus. Where Kant depicted 'taste' as an innate faculty residing in the human intellect, Bourdieu (1984) argued that taste is socially (re)produced. For the *mujeres,* the socio-political purpose of graffiti and its role in the struggle for social recognition and status for women, cholas, LGBTQ groups and others are what makes it beautiful. Theirs is an aesthetic judgement that has little to do with institutional definitions of art and more to do with its functions. Hence, the *mujeres* are explicitly concerned with how street art enables the reproduction and appropriation of political spaces; that is, how street art makes for better politics. As they have put it: *'El arte no es para los iluminados, es para iluminar! Firma: la pared'* (Art is not for the illuminated, it is to illuminate! Signed: the street).

Of the contentious performances utilised by the *mujeres*, *pintadas* have been employed with the greatest frequency. This is because the graffiti is low tech, requiring little in the way of materials and the writing style is easily replicable. As activist, Julieta Ojeda (2011) has explained: 'This cursive style is the way we were all taught to write when we were at school so it is easy for us. Any of the women can reproduce the style and that means we can spread our messages in more spaces. The consistency of style also helps people to recognise immediately that it is us speaking.' However, it is also worth noting that the collective have also designed and participated in a range of street-based enactments that have aimed to 'conquer words' (Galindo, cited by Dangl 2010). These enactments – or *acciónes*, as the mujeres call them – might be described as something between protest event and applied theatre in the sense that they use techniques of drama to address an issue of social concern. In applied street theatre, passers-by or audiences may turn into 'spect-actors' as they are prompted to use various techniques such as role-playing, improvisation and tableaux to explore challenging issues. The aim of these applied methods is to inspire critical engagement, self-reflection and even enact change. Greenwood (2012: 16) summarises well: 'learning through drama is a process that utilises the energy of the [whole] group and that develops meaning not only verbally but also viscerally, emotionally and socially. It is ... process that invites and develops the agency of its participants: an agency that includes initiating ideas, giving physical witness to those ideas and critically reflecting on those ideas in order to discard or further refine them.'

One issue that the mujeres have addressed through their *acciónes* is the rights of LGBTQ people. The Immigration and Refugee Board of Canada (1999) explain that in Bolivia, LGBTQ individuals have long faced high levels of abuse and exclusion on the basis of sexuality. Discrimination makes it difficult to gain access to welfare or specialised healthcare services. Many people feel unable to disclose their sexuality since, 'the economic and social advantages of being part of a family far outweigh the disadvantages of a gay identity, such as homophobia, concealing one's identity and leading a double life'. They continue, '[f]urthermore, the police and the courts can hardly be counted on to protect gay people, when they are threatened by problems such as violence, threats, blackmail or slander and the like'. In 1995, after a number of high-profile and extremely violent homophobic attacks in the *altiplano*, the *mujeres* used street performance to bring the image of homosexuality out of the shadows and into Bolivian daily life. In one *acción*, Julieta Paredes and María Galindo, two of the founding members of *Mujeres Creando*, painted a big red heart in the middle of a public square, using a deep red ink that was strongly reminiscent of blood. They offered flowers to spectators, drawing them into the performance and then they set down a mattress and blanket in the middle of the painted heart. Galindo and Paredes then lay down together and climbed beneath the blanket. They held their clasped hands together for all to see and then they kissed one another.[9]

Across their *acciónes* the *mujeres* attempt to modulate the symbolic and affective power of certain colours, materials, bodily gestures and objects. For instance, in one performance, the collective gathered rubbish from across the city and they placed it within the Legislative Palace. The action was at once symbolically and physically powerful. The rubbish not only signified that the governing institutions were in a state of degeneration and decay, it did so by evoking a corporeal response, that of disgust: with the stench of rotting meat and vegetables being hard to ignore. Meanwhile, in *acciónes* against the legacy of the dictatorship and other examples of violence, the collective have often used red dye to intimate the presence of blood, and crosses to symbolise death. In this way, they tap into existing cultural frames of reference and familiar visual cues and redeploy them to cast a critical commentary on the present.

Responding to the charge that the collective are more artists than activists, one member contests, 'We are street activists, we are creative women, but we are not artists and we don't want to become an artistic elite. However, we do take up our right to create and do new things.' She continues:

> Creativity is human – it belongs to all women and men. But many want to dispossess us of this creativity, something that is ours. They want to turn creativity into something elitist, saying the artists are the creative ones, the inspired ones, the ones who inspire each other. We do not allow ourselves to be dispossessed of an instrument of struggle and in everything we do, in the books we make, in the street actions, in the graffiti, we include this element

which is important and fundamental to us: creativity. Then some people say to us: 'You're artists.' But we are not artists, we are street activists. All we do is to use something which is totally human: creativity.

(Anon., cited by Ainger 2003: 258)

From rise to rupture: Bolivia's indigenous movement

As the *mujeres* became more active over the course of the 1990s, so too the indigenous movement also began a process of reawakening which has been well documented by Yashar (2005), Dangl (2010), Hammond (2011) and others. Hammond (2011) writes that Bolivia's Indian peoples had been especially vulnerable to neoliberalism's fiscal austerity, promotion of a market for land, and the growing commercial exploitation of natural resources. The displacement and relocation of ex-miners in the 1980s resulted in a doubling of the population in the Chapare and a strong reliance on coca production. Under pressure from the United States, however, President Victor Paz introduced *Ley 1008, Ley del Regimen de la Coca y Substancias Controladas* in 1988, which banned the production of coca. *The cocaleros* responded by establishing the first Andean Council for Coca Production in 1991. As a movement they grew in strength, gaining prominence within the Confederación Sindical Única de Trabajadores Campesinos de Bolivia or CSUTB, the country's largest confederation of rural workers. Neoliberal governments made various efforts to attenuate the burgeoning indigenous movement. Among other measures, the Constitution was revised in 1994 to recognise indigenous people as a group and grant them limited cultural rights. Greater administrative responsibility was also devolved to the municipal level, to give people in the local areas more power. Charles Hale (2004; 2005) writes that these measures were co-optive, representing the cultural counterpart to the neoliberal project. They intended to fashion an Indian that would substitute 'protest' for 'proposals', becoming fully conversant with modernity and posing no threat to 'the productive regime, especially those sectors most closely linked to the globalized economy'. For Hale, this 'multicultural neoliberalism' created new sociopolitical categories, namely, the *'indio permitido'* (authorised indian)[10] and his [sic] undeserving, dysfunctional Other who is 'is unruly, vindictive and conflict prone' (ibid.). Hale writes that similar attempts at political cooptation by neoliberal governments occurred across the Andean region in the 1980s and 1990s. However, in Bolivia, these endeavours ultimately backfired. Accommodations in the limited area of cultural rights failed to quell indigenous activists' demands for social justice, and the municipal reforms inadvertently increased the autonomy of *sindicatos*, many of which refashioned themselves as *ayllus* and began to field their own indigenous political candidates in greater numbers (Yashar 2005). Meanwhile, processes underway at the international level through initiatives such as the Working Group on Indigenous Populations, gave confidence and transnational support for a new wave of mobilization around the cause of indigenous rights.

In the 1997 general elections, a party called the *Asamblea por la Soberania de los Pueblos* (Assembly for the Sovereignty of the Peoples or ASP) fielded a number of indigenous candidates from the *cocalero* movement. While ASP leader, Alejo Véliz Lazo did not succeed in his bid for the national presidency, a number of others were elected to the Congress, including Evo Morales. Morales continued to grow in prominence as a vocal activist and leader among the *cocaleros* but in 2002, he was expelled from the legislature on charges of instigating violence among the *cocaleros*. Establishing his own social movement party, the *Movimiento al Socialismo–Instrumento Político por la Soberanía de los Pueblos (MAS)*, Morales continued to consolidate support while campaigning against the further liberalization of the Bolivian economy. Loayza Bueno and Datta (2011) write that support for traditional political parties fell from 70 per cent in 1991 to 38 per cent in 2005, a period characterised by extreme social and political unrest. In April of 2000 citizens of Cochabamba took to the streets to kick out foreign investors who had privatised the area's water. Meanwhile in 2003, protests erupted over the future of Bolivia's natural gas reserves. Then-president Gonzalo Sánchez de Lozada ('Goni') oversaw the violent repression of indigenous citizens who were protesting his plan to export cheap natural gas to the United States through Chilean ports. The episode sparked a popular uprising known as the 'Gas War' that led to Goni's removal from office, exacerbated disenchantment with the political class and culminated with the election of Evo Morales in 2005 as Bolivia's first indigenous president.

Supported by a highly mobilised raft of contentious actors including indigenous communities, workers, peasant and agrarian movements, Morales assumed the Presidency with the promise of redesigning the Bolivian state and its economy to redistribute wealth away from transnational elites, giving a political voice to the historically underrepresented indigenous majority and 'ruling by obeying' as he put it. In 2005, a Pact of Unity was formed between indigenous and *campesino* social movements including the CSUTCB, *Confederación Nacional de Mujeres Campesinas Indígenas Originarias de Bolivia Bartolina Sisa* (Bartolina Sisa National Confederation of Peasant, Indigenous, and Aboriginal Women of Bolivia), the *Confederación Sindical de Comunidades Interculturales de Bolivia* (Syndicalist Confederation of Intercultural Communities of Bolivia or CIDOB) and the *Consejo Nacional de Ayllus and Markas of Qullasuyu* (National Council of Ayllus and Markas of Qullasuyu or CONAMAQ). The Pact became the bedrock of support for the MAS government, particularly when it came under attack from right-wing forces (Rochlin 2007).

Notably, Morales moved to fulfil some of his campaign pledges with haste. Executive decrees and laws passed by the MAS-dominated Congress re-established state centrality in economic planning and development (Postero 2010). Morales announced plans for agrarian reform in August 2006 at the symbolic site of Ucureña, where President Víctor Paz Estensorro had signed the Agrarian Reform Act into law five decades prior. In December 2006, Morales completed a gas nationalisation programme, followed later on by moves to nationalise oil, mining, telecoms and electricity companies. Additionally, the MAS government coordinated a popularly elected Constituent Assembly to rewrite the country's constitution, incorporating the concept of 'original, indigenous, *campesino* peoples' as a reflection of the growing concord between the movements tied together in the unity pact. The new Bolivian Constitution was debated in the Constituent Assembly in 2006 and 2007 and it was adopted by referendum in 2009. The new Constitution cemented important provisions serving to recognise and assist the country's majority indigenous population, including new restrictions on land ownership and a whole chapter dedicated to indigenous rights. Additionally, it remapped autonomous governance, creating four levels of decentralised authority: the departmental, regional, municipal and indigenous. Relating importantly to the latter, the constitution also legitimated the practice of indigenous community justice and provided guarantees of prior consultation over the future development of indigenous territories in line with internationally agreed standards.[11]

Initially, Evo Morales enjoyed a high level of popularity at home and abroad. Bolivia's first indigenous president enjoyed favourable electoral results when he came to power in 2005, receiving 54 per cent of an 85 per cent turnout. Following the successful institution of many of his campaign promises, seen above, he was re-elected in 2009 with 60 per cent of the vote on a turnout of 90 per cent (Hylton 2011). As Fabricant (2011) explains, although his policies were unpopular with right wing voters, concentrated in the Eastern lowlands, Morales' credentials as an Aymara coca farmer and long-time defender of workers' rights and social justice gave him widespread acceptance from broad swathes of society, with many indigenous people identifying with him as 'one of them'. As Hylton (2011: 243) highlights, 'Regionally, Morales's Bolivia has enjoyed better relations with neighboring Chile, Argentina, and Brazil than any regime since General Banzer's, during the darkest night of Plan Condor'. In the global arena, Morales became a vocal and respected protagonist, advocating for indigenous rights and against climate change in an array of international fora. In 2007, Bolivia was the first country to sign the United Nations Declaration on the Rights of Indigenous Peoples, a document which had been 30 years in the making. The following year Morales addressed delegates at the inauguration of the United Nations' VII Indigenous Peoples' Forum where he espoused a set of ten 'commandments' for saving the planet. Moves to stage the first World Peoples' Conference on Climate Change and the Rights of Mother Earth in Tiquipaya in 2008 as well as other rhetorical stints increased Morales' popularity with leftist commentators and civil society organisations in the Global North.

Yet, for all the oratory about social justice, sustainability and environmental protection, 'there remains a considerable gap between rhetorical claims of "participatory democracy", socialism, and non-capitalist development, in contrast to the reality of policies and practices that undermine the autonomous political mobilisation, and/or economic interests, of popular sectors' (Calla and Striffler 2011: 239). A macroscopic look over the MAS period reveals deep patterns of continuity with the preceding neoliberal period. Webber (2008) points out that with the exception of moderate reforms to oil and gas policy and the extension of foreign and trade relations on the continent, the key economic imperatives of the government remain committed to a progressive development of industrial capitalism. The not-so-green extractive industries, for example, have been key priorities[12] and ministers of government have frequently referred to socialist transformation as a vague and distant objective, possible only after Bolivia undergoes a period of gradual and progressive development in its capitalist phase (Webber 2008 and Webber 2012). The government's economic vision has varying implications for the different groups that self-identify as 'indigenous' in modern Bolivia. While many of Bolivia's Indian population work in the productive sector, occupying places in the extractive industries, agriculture and services, *cholas* occupy a space in between the modern productive sector and the informal barter economy. Some smaller communities based in the Bolivian Amazon are further removed from the modern capitalist economy, limiting their productive activities to more communal and self-sustaining modes.

These divisions came to the fore in 2011, when contention erupted over the Bolivian government's plan to build a highway through the middle of the *Territorio Indígena y Parque Nacional Isiboro Secure* (Isiboro Secure National Park and Indigenous Territory, known locally as 'el TIPNIS'). TIPNIS is a national park and a self-governing territory, which covers 1,091,656 hectares and plays host to 402 species of flora and 714 species of fauna, including endangered birds and water mammals such as the pink dolphin. While providing an important source of bio-diversity, the park is also home to the estimated 12,000 people belonging to the Yucaré, Chiman and Moxeño indigenous communities, who have populated the area for thousands of years (Salgado 2011), as well as some 15,000 'colonists' or settlers who have moved into the South of the park since the 1970s. Many of the 'colonists' are former miners and they are predominantly of Aymara and Quechua origin. Unlike the Yucaré, Chiman and Moxeño communities, who operate primarily outside of the formal market economy, the 'colonists' are extractivists – including coca growers, loggers, and ranchers – and they desire increasing tracts of arable land and access to markets in order to expand production. These groups have been accused of causing deforestation, land and water pollution. Their incursion thus represents a threat to those indigenous communities whose subsistence lifestyle depends on the health and cyclic rejuvenation of the river and forest.

The proposed highway was to run from the outskirts of Cochabamba, through a pristine area of the park and jungle region to join another highway at San Ignacio de Moxos that runs from Yucumo to Trinidad. The plan was backed financially by the Brazilian government as key part of the Initiative for the Integration of the Regional Infrastructure of South America (IIRSA), with the view that it would assist in the creation of a bi-oceanic corridor, drastically reducing shipping time and costs by facilitating horizontal movement across the continent. It promised to facilitate trade links with neighbouring states and provide a direct route to from Beni to Cochabamba which would cut transportation time for perishable goods in half, while also conveniently bypassing Santa Cruz – the heart of the conservative opposition to Morales's government (Achtenberg 2011). For farmers of coca and other crops in the Chapare region of Cochabamba, market access would be improved. The gains for indigenous communities living in the TIPNIS territory were rather less clear however and in 2011, it became apparent that construction had begun both inside and outside of TIPNIS without the consultation of the indigenous communities therein. According to provisions in national and international law however, the resident indigenous groups were entitled to free, prior and informed consultation on the construction project.

Having failed to uphold these entitlements, the Bolivian government staunchly defended the road-building project, claiming that its construction would aid in the development of the indigenous communities of the TIPNIS and enable the more remote groups to access vital supplies. However, community leaders from the Yucarés, Chimanes and Moxeños found this disingenuous, pointing out that the planned route ran nowhere near the established indigenous settlements in the North of the park. Rather, the planned route ran directly through the centre-west of the national park, where virtually untouched areas of primary forest and extremely fragile eco-systems exist. The lack of consultation meant that proposed alternative routes did not get an airing and, rather than entering into dialogue with the disgruntled communities, Morales vowed that the road would be built, 'like it or not' (Achtenberg 2011).

In order to understand why the Morales government behaved in this way, we need to consider the wider politico-economic and cultural dynamics at play. As Webber (2012) explains, in terms of productive relations, the 'colonists' – those settled to the south of the park – can be considered as a 'rich' strata of the peasantry, with aspirations to expand accumulation through the appropriation of further land. As they form a key element of the MAS support base – numbering the indigenous communities of the Yucarés, Chimanes and Moxeños two-to-one – the government has been especially keen to accommodate their expansionist ambitions. However, such expansion entails political costs. Geographically, to one side of the TIPNIS 'colonists' for example, we find possibilities of expansion into the department of Santa Cruz. But, as Webber notes, 'this would imply incursions into the inhabited lands of other Aymara-Quechua migrant peasants, or into the capitalist agricultural and ranching expanses that make up the agro-industrial sector of that department' (Webber 2012). The MAS government does not want to see these communities uprooted or squeezed since they represent a crucial bulwark of support in a department that is historically more right-wing. For Morales meanwhile, the indigenous communities of the TIPNIS represent a minority of Bolivians who 'simply refuse to be modern' (Flores 2011).

Webber characterises the Morales government in terms of an 'indigenous ascendant populism': 'ascendant' in the sense of the sheer increase in indigenous public expression since 2006; and, 'populist' because the MAS regime has retained residual elements of what Charles R. Hale (2004) describes as the *indio permitido* (authorised Indian) of neoliberal multiculturalism. Indeed in the case of the *Movimiento Al Socialismo*, 'the implicit condition has evidently proven to be that very few inroads will be made on neoliberalism under its watch, provided, that is, that the popular classes and indigenous nations cannot rebuild autonomous organizational capacities outside of the governing party to force it to implement substantial reforms' (Webber 2008: 90).

The 'indigenous ascendant populism' of the Morales government is premised on an indigenism that speaks with a single voice: that of the *indio productivo* (productive Indian). The *indio productivo* is the mirror of Morales himself, a 'modern' Indian, whose lifestyle and ambition fits comfortably alongside the model of the developmental and extractive capitalist state. Meanwhile, the Yucarés, Chimanes and Moxeños – subsistence communities, engaging in more traditional communal forms of living – have been cast as 'Others' or *indios no permitidos* (unauthorised Indians) who are enemies of the state: 'unproductive', 'unruly' and 'unaware of what is in their own best interests'.

Pintando por el TIPNIS: street art in defense of the 'unauthorised Indian'

As discord around the TIPNIS issue grew during the course of 2011, a number of street artists and street art collectives emerged, rallying to the defence of the 'unauthorised Indian'. *Insurgencia Comunitaria* (Community Insurgency), friends and collaborators coordinated street art interventions in El Alto, La Paz and Cochabamba, using a combination of graffiti, murals, wheat-pastes and stencils to frame the issues, disseminate information and mobilise actors in support of TIPNIS residents. One activist articulates his motivations here: 'whilst the election of an indigenous President in 2005 represented a sea-change for Bolivian politics, Morales is not a representative figure for all Indians. He was a cocalero, so he is a colonist by background, and the policies he implements do unfortunately reflect his greater sympathies with those movements. What about the other indigenous peoples? What about the environment? Who will defend them?' (Activist, *Insurgencia Comunitaria*)

Although the $415 million contract for the highway project was signed in 2008, it attracted rather little public attention until 2011. From 2009, the most vocal opposition to the road had come from small groups of environmentalists and biologists from the state universities, who were concerned about the preservation of the habitat in the park. The very first graffiti interventions relating to the issue had appeared on the streets in 2010. In large part, these consisted of slogans, written hastily with spray paint, which took the destruction of the environment as their main theme: 'Don't destroy mother earth'; 'Forest is life' were phrases seen on the streets, particularly in the areas around the universities. However, limited coverage of the issue in the domestic and international media, meant that these phrases meant rather little to passers-by. For urban residents, it was unclear who was destroying mother earth, or why they should care. One *Insurgencia Comunitaria* activist recalls, 'since there was no news on the matter, the issue was forgotten and the graffiti, with time, disappeared'.

From June 2011 however, the issue once more came to the fore as construction began on two parts of the road. Perceiving the start of construction as an affront to their rights for consultation, approximately 1,000 indigenous TIPNIS residents and supporters responded by declaring a cross-country trek to the capital city of

La Paz to deliver their message of opposition in person. The marchers included families with children, pregnant women, youth, and elderly activists, including a 99-year-old man (Achtenberg 2011; Salgado 2011). Notably, expressions of solidarity from representative bodies in the Unity Pact, such as the *Confederación de Pueblos Indígenas de Bolivia* (Confederation of Bolivian Indigenous Peoples or CIDOB) and the *Consejo Nacional de Ayllus y Markas del Qullasuyu* (National Council of Ayllus and Markas of Qullasuyu or CONAMAQ) signalled a shift and chasm in domestic politics. CIDOB and CONAMAQ rallying against Morales on the TIPNIS issue raised the question of what it means to be 'indigenous' and who has the authority to speak on behalf of Bolivia's indigenous peoples.

Finding the governments' statements about the marchers increasingly contradictory and belligerent, a number of student and activist groups in La Paz came together in a series of open meetings to discuss ways that they could aid the marchers. One idea that came out of these initial meetings was a series of solidarity marches and demonstrations in the capital. One *Insurgencia Comunitaria* activist states that, 'during the first marches, an interesting thing happened. Taking advantage of the massive protective presence of people, some participants, including some of my own friends, just started painting. They painted on the road, on the walls... they even painted on the wall of the Brazilian embassy to express an opposition to the IIRSA project'. He continues to explain that the paintings attracted increasing numbers of amateur and professional photographers who would join in and document the demonstrations. The activists realised that creative visual strategies were a really powerful mobilising tool and resolved to continue using them to spread messages of support for the marchers.

Loose affiliations of activists acquired more concrete forms through activities coordinated around the *vigilia por el TIPNIS*. This vigil, initiated by indigenous women from the *altiplano* took the form of an occupation of the *Iglesia de San Francisco* (Church of St Francis) in La Paz. As a show of solidarity, the vigil began on 15 August, the same day as the march, and it sought to remain in place until the marchers arrived safely in the capital. One activist claims that, at the invitation of the women organisers, he and friends divided themselves into small groups to stay and sleep in the Church each night. He explains that:

> Every day the occupiers would make concerts, educational presentations and lead public debates. The idea was to create an open forum where people could get informed and participate in their own way, supporting the TIPNIS march. The vigil was the place where people could come and give donations, where reporters would come to get information about the march as it progressed. There were banners created daily and we sought to provide places where children too could come and paint. Along with that that there was plenty of graffiti being created.

In fact, vigil activities frequently spilled onto the streets, reaching larger and larger crowds of curious and receptive citizens. Displays and talks were put on with the intention of disseminating information about the communities and their legal rights as enshrined in the Constitution and international law. Some participants would stencil messages of support for the marchers around the city centre and in El Alto. These included '*TIPNIS. Carretera = Proyecto Capitalista*' (TIPNIS. Road = capitalist project). On the façade of El Alto airport, activists used graffiti to liken the MAS government initiatives to those of the Organization of American States (OAS) '*Evo – Mi heroe. Atte:OAS*'.

Others trialled more figurative interventions, playing with forms derived from life. One activist placed banners in the trees of the Sopocachi neighbourhood displaying the words '*el TIPNIS muere*' (TIPNIS dies). The sprawling red letters scrawled across the large white sheets were deliberately reminiscent of bloodstains and the positioning of the banners in amongst the trees, evoked the idea of the forest becoming a setting for violence. The jarring juxtaposition of trees and blood helped to frame the highway project as a bringer of death and destruction to TIPNIS and those who dwell there.

Another intervention in Cochabamba involved a kaleidoscope of fluorescent paper butterflies pasted against a blacked-out background. The butterflies, which provide an otherwise attractive burst of colour and cheer against a dark urban landscape, are on closer inspection, marked with death. Each butterfly displays the image of a human skull across its wingspan. Only visible up close, the skulls are an example of an aesthetic play that unsettles the familiar and surprises the viewer. While the skull is perhaps the most cogent *memento mori* in Western tradition, appearing in allegorical painting as a marker of mortality and humility (Müller-Wood 2007), for many Bolivians it has a more complex set of associations. In Aymara tradition, for example, *ñatitas* (a diminutive of 'skull') are considered as

Figure 4.5   Banner at Sopocachi, 2011.
Author's own photograph

Figure 4.6   Wheatpaste, 2011.
Photograph Courtesy of Stasiek Czaplicki Cabezas

vessels that house the souls of the dead (Nuwer 2015). Ñatitas carry an association with fertility, luck and protection. As a result, they are often kept in the home and adorned with decorations in the hope that they will intervene to bring good fortune upon the living. Notably, with Morales' rise to power, traditional Aymara practices such as the *Fiesta de las Ñatitas* (Party of the Skulls) have gained renewed prominence and inclusion as part of Bolivian national culture. They layer over and alongside the Catholic and criollo traditions that have long been regarded as official culture. Bolivia's postcolonial syncretism complicates attempts to decode some figurative street art interventions. In this case interpretations are sent in multiple directions: the skulls carried on the backs of butterflies are at once *memento mori*, totem and guardian of culture.

The art-activists reported that almost as soon as these interventions went up around the city, that there was a drive to remove them: 'the government has people who are constantly monitoring protest activity. There were teams who went around destroying all manifestations against the government'. Indeed, the government soon intensified its counter-information or 'counter-framing' campaign against the TIPNIS marchers and their broader support network. This involved a range of attempts to undermine the indigenous communities and delegitimise their claims in the eyes of the public. Since 2005, the Morales government has taken strides to control media commentary. It backed Bolivia's first state sponsored newspaper *El Cambio* and increasing numbers of journalists have been prosecuted for libel and *desacato* (or disrespect) when they have engaged in activities critical of the government (See Index 2012 and Medel 2007). On 21 August 2011, the government courted controversy by publishing the telephone records of some of the more prominent rights activists and march organisers. Among these were logs of calls received from the United States Embassy, accompanied by the accusation that marchers and NGOs were colluding with Washington in a dastardly plot to debunk Evo Morales. Increasingly, government officials used state media channels to label dissenting Indians and their supporters as 'anti-Bolivian', or as 'the puppets of imperialism'. Some activists involved with the vigil reported that they had received anonymous threats as a result.

On 25 September 2011, the Bolivian police mounted a heavy-handed intercession along the march route which dispersed marchers, separated children from parents and left some injured. Shocked and outraged by reports about the violent episode, La Paz based art-activists felt obliged to act. In various ways, activists used art as a means of 'defending' the marching Indians from further state violence. *Mujeres Creando,* for example, began to produce graffiti on the topic of government repression, inscribing the doors of female ministers of government and urging them to denounce the violence. *Insurgencia Comunitaria* also produced graffiti dealing with the topic of violence. However, they highlight that 'whilst [Mujeres Creando's] graffiti was addressed at the members of government, ours was directed at the public. We wanted to stimulate public indignation with slogans such as "Violence against children? Is this [what we want from] our government?!" We hoped that with enough public indignation, the government would be deterred from attacking them again'. Other art-activists took a different tack, using playful figurative forms drawn from the bio-diversity of the forest to boost the morale of the marchers – especially the children – and create a less belligerent atmosphere as the city prepared for their arrival:

> we made a what I will call a 'soft' intervention in the Plaza Bicentenario. It was 'soft' in the sense that we stencilled a range of 'friendly' figures from nature: river dolphins, parrots, trees, eagles and butterflies. We didn't want this intervention to shock or offend anybody, only to gently encourage the public to think about the TIPNIS and all the species that live there in harmony. The same stencils were then used to paint the floor of the *Cumbre* Pass in order to give the marchers a big welcome as they approached the city. On the whole, our forty days of solidarity and advocacy in La Paz had not been well communicated to the marchers because of the logistical challenges of relaying messages to them as they moved. However, I did see an article saying that the marchers had been happy to see our stencils along the last leg of the route. This was satisfying to me: that our art made them feel supported.

Animal stencils appeared overnight in the *Plaza Bicentenario*. They represented species found in the Amazonian reservation, such as parrots and river dolphins and they employed the colours of the wiphala flag, which has long been held as a

Figure 4.7 Amazonian Animal Stencils in the Plaza Bicentenario, 2011.
Photograph Courtesy of Stasiek Czaplicki Cabezas

Figure 4.8 Amazonian Animal Stencils on the Cumbre Pass, 2011.
Photograph Courtesy of Stasiek Czaplicki Cabezas

symbol of indigenous resistance. Once again, the authorities stepped in quickly to clear the street art. The stencils in the Plaza were cleaned up before the arrival of the marchers, but in the days to follow other activist groups spontaneously replaced and replicated them. For a time, colourful Amazon creatures proliferated across the capital, signalling the arrival of the *indio no permitido* and the first major rupture within The Unity Pact.[13]

Summary

Drawing on interview data, historical research and archival material, this chapter sought to illuminate an array of little-documented examples of political street art in Bolivia, including the muralism of Alejandro Mario Yllanes at Warisata, the counter-dictatorship movement of the *Circulo 70* and the anarcha-feminism of *Mujeres Creando*. Working its way methodically through a long twentieth century, the chapter highlighted the ways in which street art has variously responded to, mirrored and mediated political events and processes. For instance, it showed how the allegorical muralism of 'the social painters' was harnessed in the construction of the new revolutionary state after 1952, whilst the *pintadas* and other interventions of the *Mujeres Creando* enacted a 'return to the street' in the wake of neoliberal reforms in the 1990s.

The chapter highlighted how, on numerous occasions, street art has become the recourse of the 'excommunicated', providing a means for marginalised groups to gain visibility and challenge state doctrine. However as we saw, even when street art is itself sanctioned or commissioned by the state, artists themselves may deviate from the public transcript or party line. In the work of Walter Solón it is possible to see how an artist's memories, desires and dreams can become encrypted in 'official' art, making the personal and the political one and the same.

Lastly, a key task of this chapter has been to illuminate how, from *taki unkuy* and the murals of the *Escuela-Ayllu Warisata* to the graffiti of *Insurgencia Communitaria*, Bolivia's Indian population has utilised street art to frame, contest and defend worldviews, lifestyles and identities that do not align with the hegemonic model of national identity. The el TIPNIS episode in 2011 demonstrates that political street art remains an expedient tool for claim-making and political expression for Bolivia's indigenous groups, even as the country has undergone a much-lauded 'indigenous awakening'. Building on the work of Hale, Rivera and Webber, the chapter implicates Bolivia's Aymara president Evo Morales in the construction of a new 'indio permitido'. It shows how street art has been used to contest policies and pronouncements that elevate the needs of certain indigenous constituencies above others and it illuminates how art-activists have mobilised to defend the *indios no permitidos*.

Notes

1   Where MAS is a wordplay that refers to both the *Movimiento al Socialismo* (Movement towards Socialism), the political party of incumbent President Evo Morales, as well as the Spanish word for 'more'. It reads doubly as the lovers' refrain, 'I love you more' and a party-political commitment.
2   *Corregidores* (officers or magistrates) were appointed by the Spanish Crown to oversee local affairs in the colonies and strengthen Spanish authority. The *corregidores de indios* were charged with governing the Indian communities and were notoriously oppressive.
3   *Latifundios* were large estates owned by a small of wealthy patrons who were able to command the servitude of tenants in exchange for the right to farm a small tract of land.
4   The term *criollo* comes from the Spanish are verb *criar*, meaning 'to rear'. From the sixteenth century this term was used to designate the social class of those born locally whose ancestry was primarily Spanish. Under the colonial system, the *criollo* class ranked below that of Spanish born residents but above that of Indians or people of mixed descent.
5   The most infamous of these was at Catavi, a mining town near Potosi, where government troops fired into the crowd of protesters, killing anything between 19 and 700 people (Scheina 2003)
6   After ten months in power, Torres was deposed by the military and went into exile. Torres was assassinated in Argentina in 1976 in what is widely believed to have been a mission under the auspices of Operation Condor.
7   *Cholitas* are known for their colourful outfits which usually consist of a bowler hat, a long pleated skirt known as a *pollera*, multi-layered petticoats, and a thick shawl or *manta*. The outfit is an appropriation of the European style of dress that the Spanish forced upon indigenous people during the colonial period.

8 'Goni' Sanchez de Lozada played a key role as finance minister in the 1980's, heading up the NPE programme of neoliberal reform. He later served as president of Bolivia from 1993 to 1997 and was reelected in 2002, as the country again faced extreme social unrest in the face of economic distress. Goni was elected twice on an MNR ticket and remains a controversial figure in Bolivia. Heavy handed policing in the 2003 Gas War resulted in the deaths of 67 protesters and bystanders whose relatives have pursued legal action against him. Goni now lives in the United States but the Bolivian government has sought his extradition to stand a political trial for the events of 2003.

9 A video of this performance can be viewed at the NYU Hemispheric Institute's Digital Video Library: http://hidvl.nyu.edu/video/003888408.html

10 Hale borrows the phrase 'indio permitido' from the Bolivian sociologist Silvia Rivera Cusicanqui, who uttered it spontaneously, in exasperation, during a workshop on cultural rights and democratisation in Latin America. Rivera argued that the group needed to discuss how the government was using cultural rights in order to divide and domesticate indigenous movements.

11 See Bolivian Constitution Articles 30, 343, 352; the Universal Declaration on Indigenous Peoples' Rights Articles 19 and 32; the International Labour Organisation's Convention 169, Article 6.

12 Between 2005 and 2008, mineral and hydrocarbon exports rose from US$1.9 billion to $5.4 billion (Revenue Watch Institute 2012)

13 CONAMAQ and CIDOB pulled out of the Unity Pact in 2011.

References

Achtenberg, E. (2011) Road Rage and Resistance: Bolivia's TIPNIS Conflict. *NACLA.* Retrieved from: https://nacla.org/article/road-rage-and-resistance-bolivia's-tipnis-conflict

Achtenberg, E. (2013) The Enduring Legacy of Bolivia's Forgotten National Revolution. *Rebel Currents.* Retrieved from: https://nacla.org/blog/2013/4/13/enduring-legacy-bolivia's-forgotten-national-revolution

Ainger, K. (2003) Disobedience Is Happiness: the art of Mujeres Creando, *in* Notes from Nowhere (eds) *We Are Everywhere: the irresistible rise of global anticapitalism.* London: Verso. pp.256–261.

Arandia Quiroga, E. (2011) Interviewed by Ryan, H.E. in La Paz, Bolivia (14 September 2011).

Arandia Quiroga, E. (2013a) Alandia Pantoja y la revolución. La Verídica. *Otro Periodismo.* Retrieved from:www.laveridica.com/edgar-arandia-alandia-pantoja-y-la-revolucion/

Arandia Quiroga, E. (2013b) The Benemeritos' Utopia, *in* Aramayo Cruz, J. and Olivarez Rodriguez, F. (eds) *Benemeritos de la Utopia. The Aesethetics of Commitment.* Bolivia: Weinberg Press.

Balderston, D. and Schwarz, M. (2002) *Voice-Overs: Translation and Latin American Literature.* New York: State University of New York Press.

Bleiker, R. (2009) *Aesthetics and World Politics.* London: Palgrave Macmillan.

Bourdieu, P. (1984) *Distinction: A social critique of the judgement of taste.* Cambridge, MA: Harvard University Press.

Branford, B. (2004) History Echoes in the Mines of Potosí. *BBC Online.* Retrieved from: http://news.bbc.co.uk/2/hi/americas/3740134.stm

Calla, P. and Striffler, S. (2011) Reform and revolution in South America: a forum on Bolivia and Venezuela. *Dialectical Anthropology.* 35, pp.239–241.

Dangl, B. (2010) *Dancing with Dynamite: Social Movements and States in Latin America.* Oakland: AK Press.

Dear, P (2014) The Rise of the Cholitas. *BBC News.* Retrieved from: www.bbc.co.uk/news/magazine-26172313

De Oliveira Andrade, E. (2006) History, art and politics: the Bolivian muralist Miguel Alandia Pantoja. *History (Sao Paulo).* 25(2). doi:10.1590/S0101-90742006000200007

Duane Lehman, K. (1999) *Bolivia and the United States: A Limited Partnership.* Athens: University of Georgia Press.

Dunkerley, J. (1984) *Rebellion in the Veins: Political Struggle in Bolivia, 1952-1982.* London: Verso Books.

El Diario Cultural (2012) *'Círculo 70' en el Cecilio Guzmán de Rojas.* Retrieved from: www.eldiario.net/noticias/2012/2012_07/nt120709/cultural.php?n=32&-circulo-70-en-el-cecilio-guzman-de-rojas

Fabricant, N. (2011) From symbolic reforms to radical politics through crisis. *Dialectical Anthropology.* 35(3), pp.279–284.

Flores, C. (2011) A PROPÓSITO DE UNA ENTREVISTA AL PRESIDENTE DEL ESTADO PLURINACIONAL. El TIPNIS según Evo Morales . *BolPress.* Retrieved from: www.bolpress.com/art.php?Cod=2011080904). Accessed: 6 June 2012.

Fundación Solón (2011) *Walter Solón Romero, arte y compromiso.* Retrieved from: www.funsolon.org/solon.htm Accessed: 11 October 2011.

Galeano, E. (1974) *The Open Veins of Latin America. Five Centuries of the Pillage of a Continent.* 25th Anniversary edn. London: Latin America Bureau.

Galindo, M. (2010) 'They are not dignified places, hence the political issue.' María Galindo in Conversation with Alice Creischer and Max Jorge Hinderer, *in* Creischer, A., Siekmann, A. and Hinderer, M. (eds) *How Can We Sing the Song of the Lord in an Alien Land? The Potosí Principle. Colonial Image Production in the Global Economy.* Cologne: Verlag der Buchandlung Walther König, pp.55–58.

Greenwood, J. (2012) Arts-based research: Weaving magic and meaning. *International Journal of Education & the Arts.* 13(1), pp.1–20.

Hale, C. (2004) Rethinking Indigenous Politics in the Era of the 'Indio Permitido'. *NACLA.* https://nacla.org/article/rethinking-indigenous-politics-era-indio-permitido

Hale, C. (2005) Neoliberal multiculturalism: the remaking of cultural rights and racial dominance in Central America. *PoLAR.* 28(1), pp.10–19.

Hammond, J. (2011) Indigenous Community Justice in the Bolivian Constitution of 2009. *Human Rights Quarterly.* 33(3), pp.649–681.

Hite, A. and Viterna, J. (2005) Gendering Class in Latin America: How Women Effect and Experience Change in the Class Structure. *Latin American Research Review.* 4(2), pp.50–82.

Hylton, F. (2011) Old wine, new bottles: In search of dialectics. *Dialectical Anthropology.* 35(3), pp.243–247.

Hylton, F. and Thomson, S. (2007) *Revolutionary Horizons: Past and Present in Bolivian Politics.* London: Verso.

Immigration and Refugee Board of Canada (1999) Bolivia: Situation of gays and lesbians. *Office of the United Nations High Commissioner for Refugees.* Retrieved from: www.unhcr.org/refworld/country,,IRBC,,BOL,,3ae6abdd0,0.html

Index on Censorship (2012) *Bolivia Journalist sentenced to prison for defamation.* Retrieved from: www.indexoncensorship.org/2012/03/bolivia-journalist- sentenced-to-prison-for-defamation/Accessed: 1 May 2012.

Index Mundi (2012) *Bolivia Literacy Rate.* Retrieved from: www.indexmundi.com/facts/bolivia/literacy-rate

Inter-American Commission on Human Rights (1982) *RESOLUTION No 33/82, Case 7824.* Retrieved from: www.cidh.oas.org/annualrep/81.82eng/Bolivia7824.htm

Johnson, J. (2006) From Cuba to Bolivia: Guevara's Foco Theory in Practice. *Innovations.* 6, pp.26–32.

Kunzle, D. (2006) Fusion: Christ and Che, *in* Ziff, T. (ed.) *Che Guevara: Revolutionary and Icon.* New York: Abrams Image.

Larson, B. (2011) Warisata: A Historical Footnote. *ReVista: Harvard Review of Latin America.* 11(1), pp.65–68.

Loayza Bueno, R. and Datta, A. (2011) The politics of Evo Morales' rise to power in Bolivia The role of social movements and think tanks. *ODI RAPID Report.* Retrieved from: www.odi.org/sites/odi.org.uk/files/odi-assets/publications-opinion-files/7063.pdf

Lucero, J. (2008) *Struggles of Voice: The Politics of Indigenous Representation in the Andes.* Pittsburgh: University of Pittsburgh Press.

McLuhan, M. and Fiore, Q. (1967) *The Medium is the Message.* New York: McGraw Hill.

Medel, M. (2011) Journalist arrested for selling videos critical of Bolivian president. *Journalism in the Americas blog hosted by the University of Texas at Austin.* Retrieved from: https://knightcenter.utexas.edu/blog/journalist-arrested-selling- videos-critical-bolivian-president

Montoya, V. (2007) Los murales emblemáticos de un pintor revolucionario. *Rebelión.* Retrieved from: www.rebelion.org/noticia.php?id=53411

Morales, J. and Sachs, J. (1987) Bolivia's Economic Crisis. *Instituto de Investigaciones Socio-Económicas.* Documento de Trabajo No. 07/87.

Morales, W. Q. (2003) *A brief history of Bolivia.* New York: Facts on File.

Müller-Wood, A. (2007) *The Theatre of Civilized Excess: New Perspectives on Jacobean Tragedy.* New York: Rodopi.

Nash, J. (1993) *We Eat the Mines and the Mines Eat Us: Dependency and Exploitation in Bolivian Tin Mines.* New York: Columbia University Press.

Nuwer (2015) Meet the Celebrity Skulls of Bolivia's Fiesta de las Ñatitas. *Smithsonian Institute.* Retrieved from: www.smithsonianmag.com/arts-culture/meet-celebrity-skulls-bolivias-fiesta-de-las-natitas-180957289/?no-ist

Ojeda, J. (2002) Interviewed by Sophie Styles. *Z Magazine.* Retrieved from: www.zcommunications.org/mujeres-creando-by-sophie-styles

Ojeda, J. (2011) Interviewed by Ryan, H.E. in La Paz, Bolivia (17 September 2011).

Paredes, J. (2002) Interview. Translated by Pat Southorn. *Infoshop News.* Retrieved from: http://news.infoshop.org/article.php?story=02/03/18/6098398

Patch, R. (1961) *Bolivia: The Restrained Revolution.* Wisconsin: University of Wisconsin Press.

Postero, N. (2010) The Struggle to Create a Radical Democracy in Bolivia. *Latin American Research Review.* 45, pp.59–78.

Robinson, W. (2009) *Latin America and Global Capitalism: A critical Globalization Perspective.* Maryland: Johns Hopkins University Press.

Rochlin, J. (2007) Latin America's left turn and the new strategic landscape: the case of Bolivia. *Third World Quarterly.* 28(7), pp.1327–1342.

Ruiz, M. (2011) *Nuevo Mundo: Latin American Street Art.* Die Gestalten Verlag.

Sacks da Silva, Z. (2004) *The Hispanic Connection: Spanish and Spanish-American Literature in the Arts of the World.* Santa Barbara: Praeger Publishers

Salazar Mostajo, C. (1989) *Pintura boliviana del siglo XX.* La Paz: Banco Hipotecario Nacional.

Salgado, O. (2011) Bolivia: indigenous groups mobilise against highway. *Latin America Bureau.* Retrieved from: http://lab.org.uk/bolivia-indigenous-groups-mobilise-against-highway

Sanjinés (2004) *Mestizaje Upside-down: Aesthetic Politics in Modern Bolivia.* Pittsburgh: University of Pittsburgh Press.

Scheina, R. (2003) *Latin America's Wars Volume I: The Age of the Caudillo, 1791-1899.* Virginia: Potomac Books.

Seligmann, L. (1989) To Be in between: The Cholas as Market Women. *Comparative Studies in Society and History.* 31(4), pp.694–721.

Solon, P. (2007) Recuerdan a Solón como un artista de y para el pueblo. *BolPress.* Retrieved from: www.bolpress.com/art.php?Cod=2007081013

Solon, W. (n.d.) *El Quixote en la obra de Solon* Retrieved from: https://translate.google.co.uk/translate?hl=en&sl=es&tl=en&u=http%3A%2F%2Ffundacionsolon.org%2Fcategory%2Fsolon%2Fescritos-de-solon%2F&anno=

Telles, E. and Garcia, D. (2013) Mestizaje and public opinion in Latin America. *Latin American Research Review.* 48(3), pp.130–152.

Tibol, R. (1957) Miguel Alandia Pantoja: El pintor de la revolución boliviana. *ICAA.* Retrieved from: http://icaadocs.mfah.org/icaadocs/THEARCHIVE/Browse/TopicDescriptors/tabid/85/k/P/e/895011/page/21/language/en-US/Default.aspx

Walker (2008) Introduction, *in* Stavig, W. and Schmidt, E. (eds) *The Tupac Amaru and Catarista Rebellions. An Anthology of Sources.* Indiana: Hackett Publishing.

Webb, J. (2009) *Understanding Representation.* London: Sage.

Webber, J. (2008) Dynamite in the Mines and Bloody Urban Clashes: Contradiction, Conflict and the Limits of Reform in Bolivia's Movement towards Socialism. *Études socialistes.* 4(1), pp.79–117.

Webber, J. (2012) Revolution against 'progress': the TIPNIS struggle and class contradictions in Bolivia. *International Socialism.* Issue 133. Retrieved from: www.isj.org.uk/?id=780

Webber, J. (2015) Bolivia's Passive Revolution. *Jacobin Magazine.* Retrieved from: www.jacobinmag.com/2015/10/morales-bolivia-chavez-castro-mas/

Yashar, D. J. (2005). *Contesting Citizenship in Latin America: The rise of indigenous movements and the postliberal challenge.* Cambridge: Cambridge University Press.

Yashar, D (2015) Does Race Matter in Latin America? *Foreign Affairs.* Retrieved from: www.foreignaffairs.com/articles/south-america/2015-02-16/does-race-matter-latin-america

5 Argentine street art
Expression, crisis and change

Lyman Chaffee (1993: 101) states that '[p]robably in no Latin American country have graffiti, posters, and wallpaintings constituted such a popular expression as in Argentina'. Exploring articulations during the twentieth century, he concludes that in Argentina more so than other 'Hispanic' countries, street art has been deeply entwined with the development of the nation. This chapter recounts, extends and updates Chaffee's seminal work on Argentine street art. To this end, it charts the development of political street art from the early twentieth century to the start of the twenty-first century. In its stride, it takes in anti-system and pro-system interventions, including the role of street art in nation-building and examples of street art as 'infrapolitics'. It moves from a discussion of *peronism* to explore three significant street art 'interventions' or 'outpourings': the *Tucumán Arde* intervention of 1968; the '*siluetazo*' of 1983 and the stencil movement that emerged alongside the 2001 financial crisis. Based on a combination of archival research and interviews with art-activists, the chapter reveals how street art has punctuated, mediated and in some cases even spurred on processes of political change in Argentina.

From 'Golden Age' to 'the Infamous Decade': early street art manifestations in Argentina

During the period lasting roughly from 1870–1930, Argentina emerged to become a largely united, nominally democratic yet extremely prosperous and culturally acclaimed nation, its wealth surpassing that of Spain, Switzerland and Sweden. Dubbed '*El Edad de Oro*' (The Golden Age), economic growth during this period was fuelled by a variety of political and economic developments, including a relative decline in inter-state political rivalries and the disbanding of the state militias under General Roca. Following the relaxation of immigration controls, educated and highly skilled migrant workers entered in their thousands from Europe, enabling a rapid expansion of the middle class and the revitalisation of agriculture (Nouzeilles and Montaldo 2002).

As European migrant cultures merged with those of the *porteños* – natives of the port city of Buenos Aires – the popular creative enterprise of *filete porteño* was born. *Filete porteño* is a highly decorative style of painting, incorporating bright colours, a high degree of symmetry, intricate floral bordering and common motifs such as the acanthus leaf and gothic script. The *fileteado* style is now widely used to adorn all kinds of objects in Argentina and beyond. However it was first developed by Italian migrants employed in the manufacture of the horse-drawn vehicles in Buenos Aires (Genovese 2001/13; 2007; 2008) and can be considered among the earliest examples of Argentine street art.[1] Stylistically speaking, *filete* patterning is a cultural hybrid. It is strongly reminiscent of the highly decorated Roma caravans which entered use in Europe around the middle of the nineteenth century while also borrowing design elements from Italian glassware production. *Filete* can be seen as an expression of the emergent *porteño* identity – a living and evolving street-based decorative form that synthesises a diversity of aesthetic influences in an expanding multi-cultural society.

The conservative *Partido Autonomista Nacional* (National Autonomist Party or PAN) had dominated Argentine politics from the late nineteenth century. However, the massive influx of self-affirmed migrants from Western Europe in the early twentieth century brought pressure for a more open political system. Rival parties emerged to challenge what they perceived to be elitist posturing and attempts to block the emergence of competitors. Chaffee (1993) writes that at the turn of the century, the *Partido Socialista de Argentina* (Socialist party of Argentina) became the first political party to bring posters to the streets. The Socialists ran candidates for National Congress in 1896, publishing 20,000 copies of their manifesto as handbills and pasting up to 8,000 posters around public spaces in Buenos Aires (ibid.). However, a combination of electoral corruption and restrictions on suffrage made it hard for the Socialists and other emerging parties to compete effectively on the national stage (Edwards 2008).

The passing of the Sáenz Peña Law in 1912 granted universal male suffrage and required mandatory voting. It opened up the system to a range of political challengers (Alston and Gallo 2010). The *Partido Socialista*, the *Unión Cívica Radical*, UCR (the Radical Civil Union or UCR) and the *Partido Demócrata Progresista* (Party of Democratic Progressives or PDP) emerged as contenders for control of the province of Buenos Aires, all disseminating colourful campaign posters in an attempt to attract and entice voters. Chaffee (1993) notes that in 1919, Radical Party leader and '*padre de los pobres*' (father of the poor), Hipolito Yrigoyen, pushed the capacity of the new mass communication medium even further. Yrigoyen included a printed photograph of himself on his campaign materials. This was an unprecedented move and one that surprised many observers, as Yrigoyen was widely considered to be an un-photogenic man, garnering the popular nickname of '*el peludo*' (the hairy armadillo) (Blanco 2007).

Canovan (1999) defines populism as an appeal to the people, against both the established structure of power and the dominant ideas and values of a country's elite. Cammack (2000) explains that populist politics flourished in Latin America around the time of the Great Depression, as a number of charismatic, personalistic leaders appealed to broader society for support; utilising nationalistic, anti-status quo ideological framings as well as a range of patronage options. Like Vargas in Brazil, Yrigoyen sought to project an image of himself as the vanguard of the poor and labouring classes. The dissemination of his photograph on campaign posters up and down the country made him visually accessible to those who lacked consistent access to print media. Moreover, Yrigoyen's weathered and surly appearance defied attempts to frame him as a pampered and privileged political elite, lending him greater credibility with the newly enfranchised working classes as compared with many of his competitors.

Yrigoyen held the presidential office from 1916 to 1922, and again from 1928 to 1930, presiding over a number of progressive social reforms that helped boost the standard of living of Argentina's working classes. The period from the 1930s, however, began with a military coup in which Yrigoyen was removed from government and placed under house arrest. The saboteurs consisted of a group of young cadets and officers led by General José Uriburu. After removing Yrigoyen from office, Uribu's forces communicated the outcome of the *coup d'etat* to the population by distributing leaflets by airplane across the state territory. General Uriburu ruled by decree and employed torture and imprisonment against his detractors. He cancelled elections in 1931 in order to extend his rule and suppress the Radical Party, but was also hugely unpopular among his own ranks. Civilian and troop uprisings forced him to hold presidential elections in 1932, during which he was replaced by General Agustín Justo, whose supporters allied in the *Concordancia*[2] [Concordance], sought a conservative restoration and controlled democratic transition (Cavarozzi 1992). Justo instituted neither of these changes however. Instead, he moved to dismiss Congress, censor the press, purge the universities as well as declare a state of siege. In the years that followed, power moved back and forth between military and *Concordancia* governments. The period between 1930 and 1943 became known as the '*Década Infame*' (Infamous Decade), marking the end of Argentina's democratic project. The decade saw the suppression of labour groups, the return of widespread electoral fraud and a campaign of censorship and intimidation of oppositional voices.

The infamous decade saw the emergence of two of the most vocal women's suffrage movements, who in 1932 managed to get a bill to Congress providing for women's rights. The bill made it through the lower house, but its passage was thwarted by a conservative-dominated Senate (Ehrick 2015). Still held on an unequal footing to men, female members of the Argentine oligarchy began to find other ways of exercising political power, namely using charitable giving to accumulate influence and public standing. Effectively, as Vallejos (2008) points out, these women had realised that 'in the absence of voting rights they could use their family relationships, social and economic power to build a parallel power'. In this context, Sansinena de Elizalde, Maria Rosa Oliver and a young Victoria Ocampo founded the *Associacion Amigos del Arte* (AAA) in 1924. The AAA had the stated aim of improving the material welfare of local artists and facilitating access to the arts for the various levels of society. Accordingly, the Association put on conferences and talks by local and international artists, providing a space for networking and learning.

As Vallejos (2008) claims, the Association aimed to be experimental and inclusive and as a result, it tested political and aesthetic boundaries. In 1933

Victoria Ocampo invited the Mexican muralist and political activist David Siqueiros to give a series of three lectures in Buenos Aires. In the first two of these lectures, Siqueiros urged Argentines to take their art to the street and he called for the awakening of a socialist consciousness that would free creativity from the auspices of dry academicism and elite gallery circuits. Specifically, Siqueiros bid Argentina's artists to 'come out of the placid shadows of the atelier and the Montparnassian schools to walk in the full light of the human and social realities of the factories, the streets, in the working class neighborhoods, roads and huge field with its farms and ranches' (Siqueiros 1933 cited by Betta 2006, author's translation). He further claimed:

> The street spectator is distinct from the complacent gallery spectator. One has to put the painting that will be seen from afar by their eyes in such a way that the intensity of the theme and the plastic expression will be seen, will be felt. It is not possible to effect this fundamental change in painting without there being an ideological incitement. Still more: it is not possible to realize anything great without a spiritual content that encourages and strengthens this desire. We should understand that we should be tied to the grave problems of our epoch. We should lean towards the worker; we should be on the side of the weak nation pillaged by the stronger, we should hate war and aspire that artists and intellectuals enjoy greater appreciation. Art without ideological content has no reason for being and has no permanence.
>
> (Siqueiros 1933, cited and translated by Stein 1994: 87)

Siqueiros' ideas had popular uptake among the intellectual and artistic community in Buenos Aires. One inspired group petitioned General Justo to allow the Mexican artist-activist to paint an outdoor mural in the City. While Justo initially approved the idea, Siqueiros' stirring, ideological speeches had angered many on the right. The newspapers *Fronda, Cristol* and *Bandera Argentina* launched a brutal attack on his work, creating pressure for the president to withdraw support for the mural project (Stein 1994). Ocampo was drawn into the fray. Having become increasingly concerned over her own place and standing in Argentine society, she cancelled the final lecture at the *Associacion Amigos del Arte.* A political drive to get Siqueiros out of the country ensued (Betta 2006). However, before being forcibly deported in December 1933, Siqueiros managed to engage with a range of activists outside of the Argentine intelligentsia. Numerous sources attest that he left his mark – or, rather left a new technology for making marks – with labour activists in the capital. In an interview, Argentine Doctor of Letters, Claudia Kozak explains:

> The Mexican painter Siqueiros was responsible for the arrival of stencils in the country. There is a testimony about Siqueiros' visit in the 1930's. He was a part of the Communist movement and he taught the groups here in Buenos Aires how to make political stencils. Before this they had painted political paintings with tar, which was problematic because the medium was dirty and would make a mess when the rains came. Siqueiros wrote in his memoirs

about teaching these protestors how to communicate using regular masonry paint instead. The interesting thing is that Siqueiros said he used stencils. Prior to this, stencils had been used only as a mechanistic process for creating textile and wallpaper prints.

(Kozak 2011)

Siqueiros' stenciling method would prove popular in the years to come, particularly as a new workers movement emerged, taking to the streets in defence of Juan Perón.

Peronismo, street art and nation-building

Juan Domingo Perón had been one of the young cadets to lead in the 1930 coup against Yrigoyen. He rose in the ranks, becoming a colonel in 1941, Minister for Labour in 1943 and finally the elected President in 1946. The ideas and image of Juan Domingo Perón have become some of the most lasting and the most polarising in Argentine history and politics, emerging repeatedly in the country's street art. *Peronismo* is defined by some as an ideology with a core set of commitments at its base and by others as a 'brand', or even still a set of political emotions that have become an inescapable feature of national politics in Argentina (*The Economist* 2015). The rise of Perón cannot be understood without reference to practices of corruption and political exclusion during the Infamous Decade and the lasting effects of the Great Depression, which led to a revision of Argentina's economic model in favour of greater economic nationalism. The governments of the 1930s pursued industrialisation at pace and sought to reduce reliance on exports. Workers displaced from the agricultural-exporting sector arrived in the cities in droves hoping to find jobs in the new factories, contributing to the expansion of an urban working class that lived in precarious conditions and had little voice. As Labour Minister, Perón listened to these workers and instituted a number of reforms that met their needs, including improvements in labour legislation, paid leave and the creation of new syndicates for non-unionised activities. In this way he was able to cultivate a strong connection with the trade unions and workers syndicates, setting himself apart from the existing institutions.

Perón's popularity and the nature of his reforms unsettled those in power and, in 1945, he was forced to resign all posts before being imprisoned on the island of Martín García. Five days later, hundreds of thousands of workers led by the *Confederación General del Trabajo de la República Argentina* (General Confederation of Labour of the Argentine Republic or CGT) marched to the seat of government at the Casa Rosada to demand his release. Police blocked the bridges and roads to obstruct the demonstration but marchers commandeered boats to get there (Bolton 2014). The streets exploded with graffitied inscriptions and stencils supporting Perón. Leading the demonstration was Perón's then romantic partner, a popular young radio star named Maria Eva Duarte, known affectionately as 'Evita'. Such was the wave of popular support, that Perón was released. The same night, he spoke from the balcony of the Casa Rosada and declared his candidacy for the Presidency (ibid.). One month later Juan and Evita were married. Perón began his campaign with Evita at his side, the first Argentine woman to participate directly in a presidential campaign (Bolton 2014).

In response, many of Argentina's traditional parties banded together to form an alliance named the *Unión Democrática* (UD or Democratic Union). The Democratic Union fielded two of its own candidates, José Tamborini and Enrique Mosca who sought to discredit Perón on the basis of his position of neutrality in the Second World War. Then US ambassador to Argentina, Spruille Braden, had been publicly critical of Perón's position on the war, labelling him 'pro-fascist'. Braden arranged US sponsorship for the UD in the electoral race (Chaffee 1993). Angered by this perceived interference in domestic politics, Perón's supporters produced handbills and posters attacking Braden and framing him as 'the Al Capone of Buenos Aires'. Two weeks before the elections, the US State Department further conspired to discredit Perón by making public the 'Blue Book', a document detailing Argentina's and more especially Perón's dealings with the Axis powers. Rather than damage Perón's public image, however, US interference bolstered his popularity and undermined the Democratic Union. The choice for voters, as articulated by Perón, was between 'Braden o Perón'. This phrase, which painted the UD as puppets of the United States, was immediately popularised in a wave of graffiti (Chaffee 1993). Seidman (2008) recalls that Perón's campaign was also aided by a Democratic Union poster that featured him holding a white worker's vest. The image was accompanied by the slogan, 'The Sweaty One – New Colours of the Fatherland'. Embracing this defamatory image, Perón later emerged in front of crowds proclaiming that he wished only to 'mix with this sweating mass as a simple citizen!' (Perón 1945 cited by Favor 2010: 55). Holding the UD in disdain for their elitism and snobbery, the burgeoning workers' movement turned the sweating, labouring Perón into a popular icon and hero. They adopted the label '*los descamisados*' (the shirtless ones), in solidarity with him.

Winning the 1946 election in a landslide, Perón embarked on a five-step plan for industrialisation. Within this, the British-owned railway was nationalised, workers' remunerations were increased, free healthcare and education to university level were introduced. Additionally, hospitals, care homes, orphanages and schools as well as other places of refuge for the poor, sickly and dispossessed were planned and built, often under the auspices of the Eva Perón Foundation (Romero 2002). Yet, as *perónismo* became institutionalised as a hierarchically based mass party system, so too did street art shift from a largely organic form of expression to a street-based activity that was organised from the top-down. Under Perón, street art was elevated to a high symbolic level to demonstrate the grass-roots appeal of the movement. The government became a key producer of posters' and teams were mobilised by the Sub-Secretariat of Information and Press to issue a broad range of pro-Perón ephemera, including posters, placards, picture books for children, pamphlets, as well as brochures, postcards and photos of the presidential couple. Foss (2000) claims that between 1949 and mid-1951, the Sub-Secretariat produced 33 million individual items, enough to reach every man, women and child in Argentina twice over. He highlights that most of these materials were distributed for free, ensuring their wide dissemination and accessibility. The labour unions and a new *Partido Justicialista* (Justice Party or PJ) became state-appendages under the new system. The unions produced posters with common themes of social justice, nationalism and praise for Juan and Evita. The PJ meanwhile, organised in street brigades and used stencils and painted slogans to promote *peronismo* on streets up and down the country (Kozak 2011). In this period, graffiti became 'a symbol of class politics... evolving out of the opposition and protests of the working class against the dominant political elite and their culture' (Chaffee 1993: 104) and deployed in their stead.

For all his 'grass roots' appeal, Perón's style of governing was far from 'democratic'. He began his first term by dissolving and reconfiguring all of the organisations that had supported him during his election campaign (Lewis 2006). He removed Supreme Court justices and replaced them with individuals whose ideological and moral lines were drawn close to his own. Over half of the country's university professors were replaced by government-approved appointments and political activity was forbidden on the university campuses. To supplement these measures, Perón also sought to control the airwaves. Aided by loans and the assistance of influential friends, Evita purchased a range of radio networks and magazines, as well as the popular newspaper *Democracia*, which subsequently became the mouthpiece for the regime in its operations of power. Peronist forces intimidated the print media and the police were ordered to clamp down on the production of underground newspapers, pamphlets, posters and political stencils hostile to the government (Lewis 2006). In March 1949, Perón created a new constitution which permitted him to run another term in office. For a short time, public expectations converged around Juan and Evita running on the same ticket for the 1951 elections. Pro-system wall-paintings and posters echoed the expectation 'Perón-Perón 1952–58'. However this did not come to pass as Evita, suffering from cervical cancer, declined the vice-presidential nomination. She died in July 1952, casting a dark cloud over her many admirers in the Perónist movement.

Although Juan Perón was re-elected to the presidency in 1951 with two thirds of the popular vote, his electoral strategy included intimidation tactics, media censorship and risk of arrest for 'disrespect' (Lewis 2006). Notably, Perón's popularity diminished, especially after the death of Evita, and he increasingly became subject to criticism from influential actors and institutions, including the Catholic Church.[3] These strong indictments were accompanied by a worsening economic climate in which shortages of food forced people to eat black bread with millet (Romero 2002: 119). After austerity measures were introduced to help quell a growing balance of payments deficit, the labour movement turned towards opposition parties for an alternative. The government's other principal supporter, the military, became increasingly restless and in 1955, Perón was overthrown in a *coup d'etat* that cost 350 civilian lives. He was forced into exile.

From Paris to Rosario: the swinging 60s and the birth of a new aesthetic

Following the removal of Perón, Argentina experienced swings back and forth between civilian and military government. At first, the new military regime led by General Eduardo Lonardi initially resolved to establish and sustain a balance between the competing interests of the heterogeneous groups who had expressed dissatisfaction with Perón. In his view, there was great merit to be found in the nationalist-populist movement that Perón had nurtured, provided that potential agitators could be brought under control. In response to Lonardi's reasoned tolerance, many of the unions showed themselves to be conciliatory with the new military regime. However, Lonardi's unhurried approach to 'de-peronisation' angered military hardliners and business interests and he was deposed after just two months. His replacement, Lieutenant-General Pedro Aramburu, took a much more forthright approach to the peronists, immediately moving to proscribe visual symbols of the movement and issuing a decree that outlawed the mere mention of Juan Perón in public. Following the decree, 'Peronist unions were intervened, their leaderships purged. Unable to engage in electoral political activity, the workers returned to the streets to assert their symbolic power' (Chaffee 1993: 106). Graffiti was used across Buenos Aires to announce local and general strikes, demonstrations and to keep imprisoned or exiled political leaders in the public consciousness. '*Perón vive*' (Perón lives), for example, was a common refrain.

During the 1955–1972 period, a bloc of peronist unions, the *62 Organisaciones* formed a movement called *La Resistencia* who made explicit use of underground, anti-system street graphics. Levitsky (2003) outlines how the peronist movement managed to survive both within the trade unions, as well as through 'thousands of clandestine neighbourhood-based networks, or "working groups", many of which met secretly under the guise of barbeques or birthday parties. These clandestine networks organised study groups, midnight graffiti-painting brigades, masses for Evita, the circulation of Perón's messages, and literature distribution at dance halls and soccer games' (Levitsky 2003: 41–42). Not only were these infrapolitical activities crucial to the movement's post-1955 survival, they built momentum for the return of Juan Perón. When in 1964, Perón hinted that he would return from exile, *La Resistencia* plastered the walls with stencils and freehand inscriptions exclaiming '*Perón vuelve!*' (Perón returns!). However, the former leader was barred from re-entering Argentina by the military. In response, the unions launched a 'Battle Plan', carrying out work stoppages, seizing factories and 'scrawling graffiti on them symbolically to designate them as liberated zones' (Chaffee 1993: 107). Factories were adorned with banners and slogans to signal their political autonomy and freedom from the imperatives of the regime and capitalist bosses.

While they had regained strength and visibility through this combination of direct action and *infrapolitics*, the peronist factions still lacked the support of the all-important student groups and the middle classes. Things changed after 1966 when yet another military coup brought Juan Carlos Ongania to power. Ongania was an admirer of the Spanish dictator Franco. After his inauguration, he swiftly declared his intention to remain in power indefinitely and he pursued a programme of reform geared toward 'moral, political and economic revitalisation', the primary aim of which was the eradication of communism. Ongania took to purging academic, cultural and labour bodies and expressed his outright opposition to organisations that were linked to both communism and democratic reform. According to Chaffee (1993: 107) Ongania 'banned all sociopolitical activity. He intervened in the national universities using strong armed police tactics to expel students and professors suspected of "communist" sympathies, and he crushed the general strike called by the CGT'. The University of Buenos Aires (UBA) and *Instituto di Tella* – an arts centre dedicated to the development of a politically and socially engaged aesthetics – were among those targeted. During *La Noche de los Bastones Largos* (The Night of the Long Batons), students and members of staff participating in a sit-in/occupation at UBA were beaten and detained by police before being expelled from the country. Ongania's assault on the aspirations and rights of students, intellectuals, workers and others drove up support for the peronists.

Notably, 'anti-system street art dogged Ongania from the start' (Chaffee 1993: 108). Despite the regime's efforts to circumscribe political association, peronist networks expanded and their increasingly diverse composition, lack of centralisation and informality made them hard to put down. Chaffee claims that the government found itself unable to suppress street art expressions, which popped up at an unprecedented rate. Memory became a common theme in posters, stencils and graffiti inscriptions. The names of students who had been killed in clashes with the authorities were often invoked as epitaphs and indictments of government brutality. Chaffee (1993) recalls an occasion in 1966, when a student was killed during a demonstration in Córdoba. He describes how memorial posters were plastered around the zone where he was killed. Street signs were painted over by family members, friends and activists and replaced with the name of the fallen student. In this and many other examples of pop-up memorials, street art demarcated spaces *taken back* from the authoritarian state and repurposed for and by collective outpourings of anger, loss and grief.

Across the pond meanwhile, student mobilisations and strikes spread to factories and industries all over France, as part of the sequence of events known as 'May 68' or the 'Paris Spring'. Notably, *Cristianismo y Revolucion,* the dominant newspaper of the Peronist and Guevarist revolutionary left in Argentina did not mention the events in France even once. Rather, commentary centred on the so-called 'Prague Spring', with Argentina's leftist press throwing its support behind the Soviet suppression rather than the political activists (Vezzetti 2009). Despite this, alternative channels and circuits of communication brought the spirit and legacy of May 68 to Argentina, with particular effect on the artistic community, parts of which became rapidly politicised. As Kozak (2004) describes, the May 1968 protests in France became one of the most important events for the generation and evolution of Argentine street art. Nowhere is this more apparent than in the *Ciclo de Arte Experimental* (Experimental Art Cycle) and *Tucumán Arde* (Tucumán is Burning) interventions.

Greeley (2007) and others register a decisive shift not only in the look and form of art objects after 1968, but also 'in the concept of political subjectivity itself – a change in the nature of political agency' that was often actively facilitated by and through artistic practice. One of the most striking examples of this shift is in the *Grupo de Artistas de Vanguardia* who in 1968 began the conscious task of redrawing the links between artistic practice and politico-social critique. Like the Situationists in Europe,[4] they aimed to supersede the categorisation of art and culture as a domain separated from everyday life and politics. They also had a particular interest in the organisation of the built and urban environment and how it could influence the emotions and behaviours of individuals. Artist-activist Graciela Carnevale, who was part of the Rosario-based *Grupo de Artistas de Vanguardia*, directly links the group's activities to events beyond Argentina: the Vietnam War, the Cuban Revolution and the ideas emerging from Paris. She explains:

> We organised encounters and discussed the possibilities of a new aesthetics, how this new aesthetics had to be, and the new ways of the new art we were seeking, and at that time we thought the best work of art would be a work similar to a political art.
>
> (Carnevale 2006)

Carnevale and associates began to form small working groups and staged politically provocative actions and exhibitions at the Di Tella Institute, the French Embassy and other established cultural venues before moving their works closer to what they perceived to be the true site(s) of anti-government struggle: the streets and public squares.

> We... began to ask ourselves what the function of art in society was, what was the role of the artist in society. For us art did not only have to do with manual and technical skills, but was also an intellectual activity... We also wanted to address another audience, because we did not want to show our work exclusively to cultural elites, but to a wider public.
>
> (Carnevale 2006)

This process of soul-searching and critical dialogue gave birth to what the group called the 'Experimental Art Cycle'. As described in documents donated to the Essex Collection of Art from Latin America (ESCALA),[5] the 'Experimental Art Cycle' featured an array of art installations in public spaces, which sought to bridge the more revolutionary currents in Argentine politics and society with those in the Argentine art-world. The Cycle involved ten separate actions or interventions, with one taking place every 15 days. Some of these interventions played with the idea of government censorship and promoted active publics. Moreover, the cycle sought to merge life with art by 'working on the audience as the privileged material of artistic action' (Longoni cited by Bishop 2012: 118).

In this context, the first action coordinated by Norberto Julio Puzzolo seated gallery visitors on chairs facing a large window. They were required to view the street and they in turn became the spectacle of interest for passers-by. Another one of the art-activists, Eduardo Favario initiated 'Closed Gallery Piece' which used inscriptions, posters and other impromptu street signage to lead members of the public to a closed down gallery space. Once at the gallery, inscriptions at the blocked entrance re-directed people to a nearby bookshop where they were exposed to the fact that they were themselves, the work of art. As the brochure accompanying the exhibition claimed, the work affirmed the possibility of art as an 'action that tends to modify reality' by breaking down the artificial separation of art object and spectator. Moreover, in its staging of the gallery closure, it called attention to censorship and 'the impossibility that [art] might continue to develop in its traditional setting' (Favario 1968). In so doing it underlined the urgency with which to take art outside of this space, to the streets (ibid.). Other art-actions in the Cycle sought to challenge sectors of Argentine society in their passivity towards state-sanctioned violence. Rodolfo Elizalde and Emilio Ghilioni, for example, staged a street fight in which onlookers faced the moral dilemma of intervention. The fight was followed by the dissemination of leaflets and bulletins advising onlookers that they had been part of '*un arte social*' (a social art) intended to stimulate a questioning of 'ideas and attitudes that are accepted without objections out of the mere fact that they resort to authority' (Bishop 2012: 119).

Indeed, Graciela Carnevale herself sought to evoke greater sensitivities toward the routine violence committed by the police and backed by the regime. In the final action of the Cycle, Carnevale invited individuals to an opening at an empty

Norberto Julio Puzzolo
Ciclo de Arte Experimental - 1968

Eduardo Favario

Figure 5.1   Still from the Ciclo de Arte Experimental, 1968.
Photograph Courtesy of Graciela Carnevale

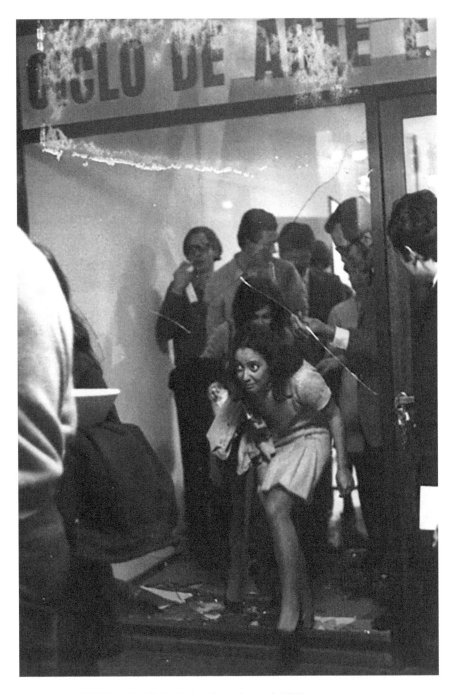

Figure 5.2   Still from the Ciclo de Arte Experimental, 1968.
Photograph Courtesy of Graciela Carnevale

storefront gallery space whose windows had been blocked out with poster art. Once everybody was inside, Carnevale left the room, locked the door, and trapped them for just over an hour until a bystander outside broke the window and allowed them to escape. In describing the work, she claimed:

> Through an act of aggression, the work intends to provoke the viewer into awareness of the power with which violence is enacted in everyday life. Daily we submit ourselves passively, out of fear or habit, or complicity to all degrees of violence, from the most subtle and degrading mental coercion from the information media and their false reporting to the most outrageous and scandalous violence exercised over the life of a student.
>
> (Carnevale 1968)

Carnevale's intervention ended in a scuffle as some of the participants blamed the passer-by for 'ruining the art'. This disturbance aroused the interest of police who used the intervention's coincidence with the one-year anniversary of Che Guevara's death as a reason to shut down the whole operation. Institutional backing from the Di Tella institute was removed. This episode marked the beginning of group's break with the formal cultural institutions. Later in the year, the art-activists met with students, sociologists and workers in Rosario and then in Buenos Aires. During these encounters, a new 'culture of subversion' was proposed, which, manifesting itself through creative initiatives, aimed to assist the working classes on their road to revolution (Padin 1997). They focused their sights on the North Eastern province of Tucumán, a key point of implementation for Ongania's national industrialisation programme.

'*Operativo Tucumán*' (Operation Tucumán), as it was called, was premised on the 'modernisation' of production and consisted in the closure of the province's traditional sugar refineries, the welcoming of transnational corporations and mass de-unionisation. Outside of the province, Ongania's media machine hailed *Operativo Tucumán* as a resounding success, giving birth to a web of myths that would conceal the region's hardship – including high unemployment, low pay, and conditions of extreme poverty and hardship – from the broader public (Albero and Stimson 1999). Having heard rumours that contradicted the official narrative about Tucumán, some of the artists, together with sociologists and journalists from Buenos Aires, visited the province to investigate the situation on the ground. They took photographs, films and conducted interviews with farmers and *campesinos*. The group documented approximately 60 thousand people unemployed, widespread malnutrition resulting in a high percentage of child mortalities, as well as the exploitative incursion of the Coca-Cola Company (Longoni 2007).

Receiving updates from their colleagues in Tucumán, an interdisciplinary group remaining in Rosario designed a counter-information campaign to illuminate the issues in the province. The first stage of the publicity campaign involved the widespread, clandestine distribution of street poster art in Rosario and Buenos Aires, which featured just the word '*Tucumán*' in high contrast black and white script. In the second stage the artists combined forces with the already active student movement who went to the streets and plazas in both cities, projecting the words '*Tucumán ARDE*' [Tucumán BURNS] on urban walls in large-scale, highly visible black masonry paint. In the following weeks, posters, graffiti and handbills declared the opening of the '*1st biennial de arte de vanguardia*' (the first biennial of art of the vanguard) invoking a Leninist tone. The 'biennial' took place across CGT headquarters in Rosario and Buenos Aires, representing a coming together of the artistic, student and labour movements (Carnevale 2016). The exhibitions themselves 'featured all-over interior environments made up of posters, placards, photomurals, newspaper montages, and an array of statistical graphs indicating rates of infant mortality, tuberculosis, illiteracy and the like, in the region of Tucumán' (Albero and Stimson 1999: xxxvi). Other components included interviews played over loud speakers, sensory manipulations in which the building would be thrown into darkness every two minutes to mark the frequency of child mortality. Juxtaposed against this information was the full range of government-sponsored misinformation (ibid.). The entire exhibition was conceived as an opportunity for revelation that would heighten public consciousness and garner media attention (Bishop 2012). Moreover, 'The movement from handbills to exhibition displays to media stratagems underscored the growing savviness of these artists to the increased role of the media in production, transmission and ultimately control of information about art and politics alike' (Albero and Stimson 1999: xxxvi).

Freeing the Argentine public from a reliance on government-sponsored reports, the Tucumán Arde intervention merged art with direct action to educate and militate against the Ongania regime. Street art featured prominently in the intervention, which in turn fed into a mounting unrest on the streets which in 1969 culminated in the *Rosariazo* and *Cordobazo* uprisings. As Chaffee writes, the escalating levels of violence and increasing politicisation were mirrored in a parallel growth in street art. 'Street barricades went up and liberated zones, heavily demarcated with graffiti, were held for hours and several days. The chaotic, war zone environment gave protestors the ability to carry out a visual assault throughout the city in a "war of the walls"' (Chaffee 1993: 108). Graffiti designated 150 square blocks of downtown Cordoba as liberated space and included slogans like: '*La obediencia empieza con la conciencia: la conciencia con la desobediencia*' (Obedience begins with consciousness: consciousness with disobedience) and '*La violencia es patrimonio de todos, la libertad solo de aquellos que luchan por ella*' (Violence is the patrimony of everyone, liberty is for those who fight for it).

Not only did the Experimental Cycle and Tucumán Arde interventions feed into a broader cycle of protest however. They also had a transformative and heuristic effect on the participating artists. Carnevale (2006) explains:

> I always say that artistic practice and my belonging to this group changed my life, my conception of the world and of art. We were young artists, although then some older artists joined too. At the beginning we just wanted to be recognised as artists, we wanted to be famous, to be invited to museums to

Figure 5.3   Tucumán Arde, 1968.
Photograph Courtesy of Graciela Carnevale

Figure 5.4   Tucumán Arde, 1968.
Photograph Courtesy of Graciela Carnevale

make exhibitions, but what we did was not traditional art, we wanted to express ourselves, we wanted to explore new languages, new materials, and we found that in our city, Rosario, this was not possible. Museums were closed to our works and experiences... We were under the dictatorship of the 60s, and it was very repressive. Not as repressive as the following dictatorship in the 70s but violence was exercised in societies in many ways, in the most intimate of ways... Young men used to have long hair. They were taken to prison and had their hair cut... we could not wear short skirts because it was not moral to do so.

Following 1968, many of the artists that had contributed to Tucumán Arde abandoned the institutional art circuit altogether, feeling an imperative to 'do more'. As Longoni (2006) puts it, the artists had staged 'a ruthless rupture with the spaces and restricted modes of circulation reserved to art – a rebellion that drove them outside or, worse still, on the opposite side of, the modernising institutional circuit with which they had shared their lives till then; a rebellion that forced them out onto the streets and made them seek alternative environments away from the field of art'. Some left the artistic field behind to take up more direct forms of activism, in some cases even taking up arms against the regime. In a 2016 interview, Graciela Carnevale recounts that one of her companions entered an armed revolutionary group and was eventually killed by the ordinary army. Others went into exile or tried to survive the political turmoil and violence that followed by going into teaching, a profession which allowed for a degree of creativity and militancy to persist in less visible ways (Carnevale 2016).

Silencio en las calles? From organised state terror to the 'Siluetazo'

Confrontations during the *Cordobazo* brought about the birth of four armed guerilla groups: The *Ejercito Revolucionario del Pueblo* (Revolutionary War of the People or ERP); the *Montoneros*; the *Fuerzas Armadas Peronistas* (Peronist Armed Forces or FAP); and, the *Fuerzas Armadas Revolucionarias* (Revolutionary Armed Forces or FAR). The latter three were mobilised around Peronist ideals while the ERP was a Trotskyist-Guevarist movement opposed to the Peronists (Chaffee 1993). These groups used a variety of strategies to disrupt and delegitimise the regime, including resorting to violence. In 1970, for example, the *Montoneros* kidnapped and killed former dictator Pedro Eugenio Aramburu. Escalating levels of violence, coupled with rising economic instability sent Ongania's approval ratings plummeting and paved the way for elections.

A brief democratic interlude allowed Juan Perón to return to the presidency in 1973. However, Perón's death in July 1974 left his third wife, Isabel to preside over a country still plagued by civil unrest, high levels of political violence and economic turmoil. Rather than tempering the aggression of armed groups, the return of Perón led to clashes between the more radical leftist and right-wing factions of the movement. Peronism became deeply divided between a youth wing, which had grown up while Perón was in exile and viewed him as a man of the left and a right wing which recalled and celebrated his corporatist and quasi-fascistic tendencies. Juan Perón and Isabel sided with the right and after Juan's death the *Montoneros* declared war on Isabel. In response, her Social Welfare Minister Jose Lopez Rega formed the Argentine Anti-Communist Alliance, which began a campaign to eliminate leftists and opponents, most notably the *Montoneros* and ERP, their friends, colleagues and families. The very first clandestine torture and detention centre (CCD) was set up in Famaillá, Tucumán, an act which can be seen as an important precursor to Argentina's infamous Dirty War. The *Operativo Independencia,* launched through Secret Decree Nr. 261 in February 1975 was signed into force by Isabel Perón, in general agreement with Lopez Rega and other ministers of government. This official document ordered and sanctioned the carrying out of military operations, of civic and psychological actions, 'in order to neutralize and/or annihilate the actions of the subversive elements in Tucumán' (Pisani and Jemio 2012:2). Isabel Perón was overthrown in a military junta in 1976 and replaced by General Jorge Rafael Videla, setting into motion *El Proceso del Reorganización Nacional* (The Process of National Reorganization', also ominously termed 'The Process') during which up to 30,000 individuals were forcibly 'disappeared'.

The full extent of atrocities committed by the military government, through the so-called Task Groups, which largely consisted of junior military officers, non-commissioned officers, off-duty policemen and committed civilians began to emerge only after the fall of the regime when the *Comisión Nacional sobre la Desaparición de Personas* (National Commission of Disappeared Persons, or CONADEP) report was made public and members of the military establishment began to be called to trial for their crimes. Many years on, it is understood that 'the process' was divided into four determinate events or moments: kidnapping, torture, arrest, and execution (Romero 2002), all of which exhibited profound levels of cruelty and psychological divisiveness:

> For the abductions, each group organized for that purpose—commonly known as "the gang" (*la patota*)—preferred to operate at night, to arrive at the victims' homes, with the family as witnesses; in many cases, family members became victims themselves in the operation. But many arrests also occurred in factories or workplaces or in the street, and sometimes in neighboring countries, with the collaboration of local authorities. Such operations were realized in unmarked but well-recognized cars— the ominous green Ford Falcons were the favorite—a lavish display of men and arms, combining anonymity with ostentation, all of which heightened the desired terrorizing effect. The kidnapping was followed by ransacking the home, a practice that was subsequently refined so that the victims were forced to surrender their furniture and other possessions, which became the booty of the horrendous operation.

> (Romero 2002: 217)

Videla's forces utilised a combination of coercive and psychological tools to invoke and manage a system of 'politically-determined fear'. Those taken away by the Task Groups were usually subjected to prolonged periods of torture in which they might be subjected to sexual abuse; simulated drowning; electrocution, as well as forms of psychological abuse including having to witness the torture and killing of loved ones. These methods were carried out under the premise of information extraction, identification and location of dissenters. However, as Romero (2002: 217) notes, 'generally it served to break the resistance of the abducted persons, to annul their defences, to destroy their dignity and personality'.

The green Ford Falcons were but one signal of the regime's covert violence and total impunity. The growing absence of oppositional inscriptions in public spaces and a preponderance of disinformation through conventional media channels and pro-system street art were others. Chaffee (1993) documents the first six months of Videla's regime, relaying that initially street art was vigorously produced by the four guerrilla groups. However, as the regime became increasingly repressive, the visibility of oppositional graffiti and other street art forms receded. The regime, however, conspired to produce its own street art posters and graffiti in order to project the illusion of public support and eliminate any imaginary of the street as a space for opposition and collectivity.

> There were many victims, but the true objective was to reach the living, the whole of society that, before undertaking a total transformation, had to be controlled and dominated by terror and by language. The state became divided in two. One-half, practicing terrorism and operating clandestinely, unleashed an indiscriminate repression free from any accountability. The other, public and justifying its authority in laws that it had enacted, silenced all other voices. Not only did the country's political institutions disappear, but the dictatorship also shut off in authoritarian fashion the free play of ideas, indeed their very expression. The parties and all political activity were prohibited, as were the labor movement and trade-union activity. The press was subject to an explicit censorship that prevented any mention of state terrorism and its victims. Artists and intellectuals were watched over. Only the voice of the state remained, addressing itself to an atomized collection of inhabitants.
>
> (Romero 2002: 219)

Although some clandestine activities continued, the sheer terror and brutality of life under the regime led to the waning of direct action. One of the few signs of public defiance and opposition came in the unlikely form of the *Madres y Abuelas de la Plaza de Mayo* – a group of mothers and grandmothers who began a weekly march and vigil on the capital's central square, silently demanding the return of their children and grandchildren who had been disappeared by the regime. The *Madres* began as individuals searching for their children through legal means in government offices and then realising their lack of progress, came together, using what Chaffee would term 'auxiliary street art modes' in order to identify one another and communicate their aims to the Argentine public:

Our first problem was how we were going to organise meetings if we didn't know each other. There were so many police and security men everywhere that you never knew who was standing next to you. It was very dangerous. So we carried different things so we could identify each other. For example one would hold a twig in her hand, one might carry a small purse instead of a handbag, one would pin a leaf to her lapel, anything to let us know this was a Mother... we tried to produce leaflets as well – we had to do it secretly because it was illegal of course – and little stickers saying the mothers will be in such a place on such and such a day and 'Donde estan nuestros hijos desaparecidos?' [Where are our disappeared children?] or 'Los militares se he llevado nuestros hijos' [The military have taken our children. We went out at night to stick them on buses and underground trains. And we wrote messages on peso notes so that as many people as possible would see them. This was the only way to let people know that our children had been taken, and what the military government was doing, because when you told them, they always said 'They must have done something'. There was nothing in the newspapers; if a journalist reported us, he disappeared; the television and radio were completely under military control, so people weren't conscious. In the beginning we had no support at all.

(Dora de Bazze cited by Fisher 1989: 53)

However, these acts of 'infrapolitics' were not enough for the *Madres*. Where countless others had been deterred from confronting the regime so directly, the *Madres* decision to install a permanent weekly presence in the Plaza de Mayo was 'an act of desperation rather than one of calculated political defiance. It was an act of desperation which the mothers believed only other mothers who had lost their children could share' (Fisher 1989: 52). As one of the *Madres*, Maria del Rosario explains, that this presence manifested as a silent march was something of an accident. The women arrived at the square aiming to talk about their disappeared relatives and knit. Yet, whenever more than two of the *Madres* sat down together, they were speedily moved on by gun-wielding military personnel (ibid.). As a result, the women took to circling the Plaza on foot, walking two by two. This silent march became a weekly fixture. In order to recognise each other and heighten their visual impact, the women wore white kerchiefs tied over their heads. Each kerchief was embroidered in coloured thread to display the name of the woman's disappeared kin (Chaffee 1993). Sometimes the women would carry photographs of their missing loved ones around their necks or raised on banners circling the obelisk in a kind of mobile collage.

Initially, the military paid little heed to the *Madres* and *Abuelas*, not considering for a moment that a group of animated older women could pose a credible threat to their vast machinery of intimidation, disinformation and control. The regime used the mainstream media to deride the women as '*las locas*' (crazy women) but, as Fisher (1989) explains, the *Madres* confronted the military with an image that they struggled to discredit. The 'silent, accusing presence of the mothers without their children, implied that it was the military who had damaged stable family life and undermined Christian values; it was military who lacked legitimacy' (Fisher 1989: 60). As the movement grew and gathered international attention, the junta cracked down and, at the end of 1977, 14 of the mothers were themselves disappeared.

The 1978 World Cup was hosted by Argentina and the junta saw this as an opportunity to rally national pride. Fisher (1989: 72) notes that, 'while the foreign television cameras were focused on Argentina, the junta would put into operation all the propaganda machinery at its disposal, to create an image of peace and stability'. However, the presence of the international press also provided the *Madres* with an opportunity to be seen and heard by a broader transnational constituency. Airtime with international broadcasters helped to connect the *Madres* 'with members of governments and progressive social movements across Europe' (Bosco 2001: 319), which in turn gave them access to rhetorical and practical support including legal expertise and monetary assistance. As a result, international human rights observers, church groups and a specially convened United Nations Committee on Involuntary Disappearances began to press the junta for information and transparency. This is a particularly good example of what Keck and Sikkink (1998) have referred to as 'the boomerang effect': when activists are able to circumvent the indifference or resistance of state actors at the domestic level by transferring debate to the international level and securing pressure for change from outside. Hence, when US President Jimmy Carter suspended aid to Argentina, the junta, desperate to repair their international image, invited in observers from the Inter-American Human Rights Commission. The report, released in 1979, was fairly damning and the military regime responded with a kind of hostile desperation, disputing the Commission's findings and producing series of posters that proclaimed, 'We are right and human' (Chaffee 1993).

Veigel (2009) directs attention towards a multiplicity of factors, which contributed directly or indirectly to the total breakdown of the military dictatorship after 1982. These included increased attention to the atrocities by the international community – thanks in large part to the activism of the *Madres* and *Abuelas* of the Plaza de Mayo, economic instability and finally the junta's defeat in a 74-day war with Britain over the Malvinas/Falkland Islands. In its very first public critique of the regime, the Conference of Argentine Bishops of the Roman Catholic Church issued a document in May 1981 that questioned the economic policy and methods used in the regime's war against subversion (International Center on Nonviolent Conflict 2010). In early 1982, President Leopoldo Galtieri, then head of the military junta, authorised the invasion of the Malvinas/Falkland Islands. The operation was designed to draw attention away from human rights and economic issues at home by bolstering national pride and giving teeth to the nation's long-held claim on the islands. During the war, the regime sponsored a raft of posters and murals that targeted British Prime Minister Margaret Thatcher as the enemy, depicted the islands in the Argentine national colours of white and blue and commemorated the dead as martyrs to nationalism (Chaffee 1993). However, as the extensive costs of the war became apparent, the walls castigated Galtieri and the regime in 'a stream of anti-armed forces sentiments' (ibid. 111).

The closing stages of the dictatorship and reawakening of civil society were signalled in an outpouring of street art, which amplified opposition to the repressive state machinery and called for the *reaparición* (reappearance) of the disappeared. In 1983, the *Madres* and collaborators led in one particularly jarring street art intervention which brought hundreds of people together in the streets to produce life-sized silhouettes of the disappeared. Drawn or printed on paper and then pasted 'in a standing posture, onto walls, trees, and pillars around the Plaza de Mayo' (Longoni 2007), these figures represented the 'missing': family members, friends, community leaders, teachers and students. Invoking the memory of past civil uprisings, this street art event came to be known as the *Siluetazo*. Beginning in September 1983, and coinciding with the *III Marcha de la Resistencia* (Third March of Resistance), the *Siluetazo* was initiated by the *Madres* of the *Plaza de Mayo* in collaboration with three visual artists: Rodolfo Aguerreberry, Julio Flores and Guillermo Kexel.

Aguerreberry, Flores and Kexel had taken inspiration from the work of Polish artist, Jerzy Skapski, whose visual commentary on Nazi concentration camps had been featured in a 1978 volume of the United Nations Educational, Scientific, and Cultural Organization (UNESCO) magazine, *The UNESCO Courier* (Longoni and Bruzzone 2008). On page 22 of the magazine, Skapski's poster featuring 24 rows of differentiated silhouettes (2,370 figures in total), sat above the following caption:

> Every day at Auschwitz brought death to 2,370 people, and this is the number of figures represented above. The concentration camp at Auschwitz was in existence for 1,688 days and this is the exact number of copies of this poster printed. Altogether some four million people died at the camp.

In a letter to the magazine, Skapski wrote: 'When I had finished painting this poster I was afraid to put my name on it. What meaning have names in comparison with people's lives?'. Like Skapski, the artists wanted to create a visual representation that would call attention to the staggering number of bodies disappeared by the regime. Julio Flores expands: 'The objectives were to reclaim through representation, the lives of the disappeared ... to create a graphic that would shock the Government through its physical scale and its formal development, and to create something so unusual as to renew the attention of the media and cause a provocation that would last many days before leaving the streets' (Flores 2004, cited by the Museo de Arte Contemporáneo de Rosario 2009). Initially the artists had hoped to submit their idea for an exhibition prize but failing to find an institution willing to take on a work which called out the military – then, still in power – in such dramatic fashion, they turned to the *Madres* of the Plaza de Mayo for assistance.

The *Madres* were in favour of taking this intervention to the streets but they enforced certain aesthetic and symbolic principles of their own. Firstly, the aim of the silhouettes would be to call for reappearance rather than mourning. This was broadly in line with the *Madres* call for '*aparece con vida*' (appearance with life).

Since the regime refused to release any information on the whereabouts of the 'disappeared', there was hope that they were still alive walled up in a prison camp somewhere. Activists called for the disappeared to be returned as they were taken, that is, alive. In order assuage the temptation of associating the silhouettes with death, the mothers insisted that the silhouettes should be pasted upright rather than on the ground. They pre-prepared a large number of figures in order to impose a certain aesthetic uniformity. The idea was that these silhouettes might represent any one of the disappeared – fathers, sons, daughters, friends, neighbours (Longoni 2007).

On 21 September, the artists and the *Madres* descended on the *Plaza de Mayo* armed with rolls of wood paper, 1,500 pre-prepared silhouettes, paints and stencils. They began pasting up the life-sized figures, hoping that others would join them. The response far exceeded their expectations. Within hours the Plaza had transformed into a gigantic public workshop for the production of silhouettes. Initial plans to create uniform bodies were crushed by the spontaneous initiatives of passers-by who adapted pre-prepared figures or traced new ones in order to better represent the physical traits of their own disappeared relatives. Some silhouettes had glasses or beards, others had the spinal curvatures of the elderly or the rounded bellies of expectant mothers. Children lent their bodies to be traced in representation of missing youngsters. The artists reflect that the extent of public involvement soon made them dispensable: 'I think within half an hour of reaching there we could have left the Plaza because we were not needed for anything' (Aguerreberry cited by Longoni 2007: 180).

The first *Siluetazo* recaptured the public space of the *Plaza de Mayo* so powerfully that the action was repeated across other locations in Buenos Aires in the months to follow. Longoni (2006: 3) summarises, the *Siluetazo's* 'remarkable impact was due not only to its mode of production (the demonstrators lent their bodies for hundreds of artists to outline their contours, which in turn came to stand for each of the disappeared) but also to the effect achieved by the crowd of silhouettes whose voiceless screams addressed passers-by from the walls of downtown buildings the following morning'. In these ways the event ceased to be a purely artistic endeavour but was rather a socialisation of creative production – an action that came to occupy a space on the boundary between art and politics. The artist Leon Ferrari who participated in the *Siluetazo* reflects that the event was 'formidable not only politically but also aesthetically. The number of elements that went into play: an idea proposed by artists, carried out by the masses without any artistic intention. It is not as if we got together for a performance, no. We were not representing anything. It was a production of what everybody felt, whose material was inside the people. It did not matter if it was art or not' (Ferrari cited by Longoni 2007).

As Ferrari's comments indicate, the *Siluetazo* was a perfect storm of the strategic and aesthetic. Beginning with a carefully planned protest action, silhouette production soon took on a life and course of its own. Moreover, in as far as the action was carried out by the masses as an expression of 'what everybody felt' the event had a cathartic dimension. The *Siluetazo* went a long way towards breaking

Figure 5.5 The *Siluetazo*,1983.
Photograph Courtesy of Monica Hasenberg

Figure 5.6   The *Siluetazo*, 1983.
Photograph Courtesy of Monica Hasenberg

the pact of silent complicity with the regime, encouraging a widespread public acknowledgement of the atrocities of the Dirty War. Moreover, the event itself marked a sudden and dramatic turning point in public perceptions about the political environment, the balance of opportunities and threats for activism. The sudden outpouring of anti-regime sentiment, tipped the general mood of suspicion and fear amongst *porteños*. The visual materialisation of so many bodies – of paper and flesh – on the streets, signalled a new confidence among the populace and an end to the political and artistic asphyxia which had characterised the years of the Dirty War.

Street art, democracy and justice?

On October 30 Argentines went to the polls to elect a president; vice-president; as well as national, provincial and local officials. The fall of the regime at the end of 1983 brought with it an outpouring of street art. New and old political players were represented in a burst of visual noise. Posters, handbills and graffiti adorned walls, windows, and shutters around the city of Buenos Aires. Much of it condemned the Dirty War and expressed hopes that the new Argentina would be 'for democracy – against coups'. Artists like Ral Veroni began to develop techniques for diffusing their art to wider publics. Veroni identifies these as 'a poem written on paper and photocopied, then pasted to walls, screenprinting applied to non traditional mediums like paper money that had gone out of circulation, and stickers... Stickers for intervening in political posters, graffiti, stencils' (Veroni n.d.). Political groups that had been badly repressed by the regime recovered, some faster than others. New groups, new alliances, and splinters from old collectives formed, broke apart and reformed 'in a continual struggle for survival, recognition and greater leverage' (Chaffee 1993: 116). In this process, 'each group used street art to advertise its metamorphosis, its identity and its realignment' (ibid.) as well as to make demands on the new government headed by the *Unión Cívica Radical's* Raúl Alfonsín.

Several key themes or motifs emerged in the street art interventions made by political parties in the wake of the transition. These included: national and regional elections; accountability for human rights abuses[6]; international solidarity with the *Sandinistas* in Nicaragua[7] and notably, the direction of macroeconomic policy. In an attempt to stabilise the economy, Alfonsín borrowed from the IMF and followed its dictates. He imposed price controls which were negotiated with 53 leading companies and he also froze the wages of state workers. These choices earned Alfonsín a bad reputation with those on the left. In a vivid reawakening, the peronists produced street art posters framing Alfonsín as a 'Coca Cola candidate' and puppet of North American capitalist imperialism. Meanwhile, the trade unions and workers inscribed the walls with '*Minga al FMI*' (Screw the IMF) and demanded 'popular consultation on the international debt. One CGT poster showed a starving child and stated 'do not pay the international debt this way – only with social justice will democracy be consolidated'.

The promise of democracy dwindled under the weight of neoliberal measures that called for a total re-conceptualisation of the state in its interactions with civil society. Neoliberal restructuring began under Alfonsín and then accelerated under the Presidency of Carlos Menem. Salaries were slashed, national enterprises sold off and public sector employees were laid off. Menem's restructuring included as its centrepiece, the 1991 Convertibility Law, which pegged the peso to the US dollar. Heralded as a neoliberal success story, Argentina exhibited strong economic growth during the 1990s. Speculative FDI poured into the country and capital became increasingly concentrated in large transnational firms and privately owned banks. However all the while, the fixed exchange rate reduced the demand for exports. Production stagnated, with knock-on effects for levels of formal employment. The informal economy expanded, exacerbating inequalities, reducing tax revenues and producing persistent and dramatic increases in the federal deficit.

Over the same period, the governments of Alfonsín and Menem also attracted criticism for oscillating back and forth over their human rights obligations. Raul Alfonsín, the first democratically elected president after the end of military rule, 'inherited a weakened democratic infrastructure and a strong military that actively resisted accountability for past crimes, frustrating initial justice efforts' (International Center for Transitional Justice 2005). Alfonsín set up CONADEP, charged with investigating the fate of the disappeared. In 1984, CONADEP released the report, *Nunca Más* (Never Again), which listed detention centres where tens of thousands of individuals were murdered and/or tortured. Although Alfonsín's government successfully initiated proceedings against the military, it also passed two laws that curbed prosecutions. *Ley de punto final* (Final Point Law) established a limited timeframe for making criminal charges against alleged human rights violators, while *Ley de Obediencia Debida* (Due Obedience law) expressed that soldiers and police who were following orders from a higher authority could not be held legally responsible for their crimes. Therefore, only the highest leaders – including two former military presidents, Jorge Videla and Roberto Viola, former navy commander Emilio Massera, and former army generals and Buenos Aires Police chiefs Ramon Camps and Pablo Ricchieri – were initially tried. All of them were later pardoned by Carlos Menem, in a controversial move aimed at 'closing a decisive chapter of the past' (Marx 1990).

Menem's decision came as a blow to victims and their families, and it foreclosed on many options to continue pursuing justice for past crimes. Over the course of the 1990s, political street art became a tool and remedy for groups seeking truth, justice or closure related to crimes committed by the junta. The *Madres y Abuelas de la Plaza de Mayo* continued to stage marches demanding clarification of the fate and whereabouts of the disappeared and new groups rallied to their side. In June 1996, the *Madres'* thousandth weekly march, was marked by the presence of a large contingent of children of the disappeared. The *Hijos e Hijas por la Identidad y la Justicia contra el Olvido y el Silencio* (Sons and Daughters for Identity and Justice Against Oblivion and Silence or HIJOS) walked well behind their elders, in a way that called attention to the missing generation between them (Collard 2012). In collaboration with other new art-activist groups, particularly the *Grupo del Arte Callejero* (GAC) and *Etcetera*, *HIJOS* developed and popularised a performative and highly effective method of protest, known as *escrache*.

Derived from the verb *escrachar*, an Argentine slang term meaning roughly 'to uncover', *escraches* have been described as 'something between a march, an action or happening, and a public shaming' (Whitener 2009: 21) *Escraches* emerged at a juncture when criminal proceedings had stalled. Many former torturers or '*represores*' (repressors) had been 'recycled' through private security companies, taken on new identities and been enabled to live with impunity in the post-dictatorial context (Longoni 2008). *Escraches* were developed as a means of disrupting this reality and forcing *represores* to face up to the consequences of their actions. By exposing the *represores*' past ill-deeds to his/her co-workers, neighbours, friends and family, the *escrache* ushers in a process of reflection, reproach and castigation by the community.

Escraches are extremely well organised. About a month before each planned event, HIJOS prepare the surrounding community. 'They work with neighborhood organisations and go door-to-door to discuss with individual residents and families what the person did and the need for denouncing it' Whitener (2009: 20–21). Next comes the information campaign, which usually involves a variety of visual and spatial interventions that call attention to the sites of clandestine detention centres, extrajudicial killings, disappearances and – importantly – the new residence of the accused. Since 1998, GAC have been supporting *escraches* by producing and deploying replica road signs, which provide alternate maps of Argentina's socio-historical space. Meanwhile, *Etcetera's* main contribution has been through their striking examples of street theatre, wherein huge mannequins or masked characters play out scenes of torture, kidnapping and confession, invoking the real or imagined experiences of victims, in all their barbarity. The intention here is to shock and unnerve through a public spectacle that draws on a repository of socially reproduced meanings and symbols in a way that recasts power relations in favour of the victims.

Collard (2012) explains that, '[w]hen the time for the escrache arrives, many neighbors enthusiastically join the demonstrations or wave from their balconies, looking down at the massive spectacle'. However, others retreat and observe from inside their houses. She asks whether this latter reaction is due to opposition, indifference or rather, fear. As she points out, some *escraches* have been violently suppressed by former criminals who have received police support, but 'this is an unmistakable sign of both their impact and the continuing loyalties and connections the represores could rely upon within the police force' (Collard 2012).

Escraches at once combine political street art with a kind of 'Do It Yourself' justice. In one of HIJOS earliest published statements they express 'there is escrache because there is no justice' (Whitener 2009). *Escrache* and its attendant street art forms thus arguably fulfilled an important social and judicial function in the absence of effective institutional channels. Yet in Argentina and elsewhere, they have divided public and scholarly opinion. While some have acknowledged the role of the Argentine *escrache*s as a catalyst ushering progress toward criminal trials for torturers, which resumed in the late 1990s, others have been rather more wary of the ways that these direct actions allow protestors to take matters into their own hands without fair trial or due process. As such, some have likened the Argentine *escrache* to a form of mob justice and vigilantism that has the potential to engender and escalate cycles of violence and retribution.

Escraches are not the only form of urban intervention that combine street art, community action and calls for truth about past atrocities. Other civic groups have etched their own sites of memory and trauma into the city's material landscape. One example of this is the colourful *baldosas* (tiles), which can today be found in over 200 different locations across the capital city of Buenos Aires (Castro 2014). An initiative of the community group *Barrios x Memoria y Justicia* (Neighbourhoods for Memory and Justice), the *baldosas* mark the locations where disappeared citizens were last seen by their neighbours, family or friends. 'By mapping these significant and traumatic places in the city, the past becomes present and ordinary spaces that we pass daily carry a different meaning. No longer only for those who were already aware of its history, but also for those who were not. By giving memories a physical place in the city, personal and specific events become part of a bigger story, of the collective memory and of the cityscape of Buenos Aires' (Dirks and Siemerink 2013). But not only do the tiles record the past. The making of *baldosas* has become a ritual in itself. Often a small group of family and friends will meet to create the tile together, sharing feelings and collectively reconstructing a memory of the disappeared. Vincent Druliolle (2011) describes such micro-memory projects as deeply political, underlining the violence with which silence and mistrust were enforced by the junta. By performing

Figure 5.7   A 'baldosa' in Buenos Aires, 2011.
Author's own photograph

the task of remembering through collective art production, *barrios* enact a form of civics that breaks down barriers, foments trust and rebuilds social bonds.

'Que se vayan todos': street art in crisis

If one of the most important inspirations or catalysts for the production of political street art in Argentina's post-dictatorial period has been the memory of violence and state-sponsored terror, then the other has surely been economic crisis. The spectacular collapse of the Argentine economy in 2001 has been the topic of much past debate by local and international economists, political scientists, politicians and sociologists. A decade of neoliberal programming that reduced welfare, damaged local industries and made the economy vulnerable to speculative flows has been blamed, as have corruption and economic mis-management, including the sustained reliance on an artificial exchange rate, embodied in the 'convertibility system'. While it is impossible to isolate a single explanatory factor for the crisis, the government's ill-advised actions in the wake of extreme hyper-inflation exacerbated economic and social woes. December of 2001 began with a freeze on banking transactions which became known as the *corralito* (little corral). As a result, the economy turned from recession to depression as people and businesses could not make payments. Credit evaporated and tensions finally spilled over. Argentina's middle classes, denied access to their dollar-denominated savings, took to the streets alongside an outraged proletariat, banging their empty pots and pans and demanding *'que se vayan todos'* [throw them all out]. Systemic shock – both economic and social – led to a widespread indictment of the entire political class.

As anger and frustration mounted, the walls of the capital city of Buenos Aires became flooded with graffiti, posters, banners and other seemingly spontaneous outbursts – the scale of which had not been seen since the fall of the regime almost 20 years earlier. Common refrains included the now emblematic *'que se vayan todos'* [throw them all out] *'violencia es robar'* [It is violence to rob] and *'congreso traidor'* [traitor congress]. Other inscriptions recalled previous failures of government to uphold the social contract. Perhaps most poignantly, protesters invoked the atrocities of the Dirty War with allusions to '1976' – the year in which *El Proceso del Reorganizacion Nacional* began. The phrase *'nunca más'* [never again] – adopted from the title of the CONADEP report into disappearances under the junta – was liberally scrawled across walls and adapted to fit the context: *'nunca más bancos'* [banks, never again].

Joining this visual mêlée were a 'striking number of groups composed of visual artists, film and video-makers, poets, alternative journalists, thinkers, and social activists' (Longoni 2006: 4) who trialled new ways of communicating and organising social life against the backdrop of the broken political system. 'These new ways comprised popular assemblies, pickets and factories recovered from inactivity by their former workers, movements gathering the unemployed, bartering clubs, etc.' In sum, hundreds of thousands of people became engaged in autonomous self-organised projects or *autogestión*, finding ways to solve problems collectively and horizontally (Sitrin 2012).

The *Taller Popular de Serigrafía* (Popular Workshop for Serigraphy/ Silkscreen), for example, utilised a form of mobile silkscreen printing to produce posters that called the population out to the streets to demonstrate. They also printed and distributed T-shirts, handkerchiefs, banners, sweatshirts: whatever people could wear and then 'take off' in amorous demand' for political change (Longoni 2008: 4). Meanwhile, a number of stencil collectives also emerged in this period, updating the long valorised technique of political stencilling to provide their own unique forms of commentary on the unfolding context. Journalists and scholars have corralled these collectives and their interventions under the heading of a 'grassroots democratic movement' (see, for example, Lyle 2007). However, many of them describe themselves as neither as 'activists' nor as 'artists'. One member of the collective BSASSTNCL claims: 'None of us were political activists. None of us had ever painted in the streets...' Meanwhile, one member of StencilLand emphasises that: 'I do not think that the intention of those who painted in the crisis and explosion of 2001 was art. It was a generalised reaction.' Rather than an identify an explicit political aim, ideology or strategy, the stencil practitioners variously allude to an engulfing energy, mood and compulsion that drew them out to the street:

> It was in the air... You would see all the people in the streets and think, 'I just have to do something.'
>
> (GG, BSASSTNCL 2011)

> I lived downtown and everything was happening all around me... the City was in the mood. It was hot and no one had any money
>
> (NN cited by Lyle 2007).

> In 2002 I started seeing on my way to work, one stencil, the next day another, and with each passing day it looked like the walls were made of mutating colours... I asked myself why people would do this [spending time and money painting pictures on the walls of the city]...Without ever finding the answer to that question, I cut a stencil, I bought a spray can, and that night I went out to paint.
>
> (StencilLand 2011)

Moreover, interviews with the stencil collectives, BSASSTNCL, StencilLand and Vomito Attack, reveal some of the unique and under-acknowledged ways that street art does political work in the midst of crisis, for example through aesthetic play. The classic play theory of art was expounded by Friedrich Schiller in his *Letters on the Aesthetic Education of Man*. Schiller contended that human and non-human animals possess a primary 'play impulse' which is at once intrinsically and extrinsically valuable. On the whole, the concept of play remains profoundly under-theorised in the social and political sciences. However, Hein (1968), Latta (2002), Upton (2015) and others, working from distinct disciplinary backgrounds in philosophy, computer gaming and management theory draw on the work of

Schiller to highlight how meanings are remade, and how boundaries and routines may be questioned through examples of play.

Forms of play are exhibited in the work of two of Buenos Aires' most well-known stencil collectives, BSASSNTCL and StencilLand. Some of BSASSNTCL's earliest interventions made use of experimental juxtapositions, humour, irony and exaggeration to expose and criticise everyday obstructions, annoyances and fractures in Argentine political culture and society. Many had the effect of invoking a self-reflexivity by encouraging citizens to reflect on their own powerlessness and lack of agency vis-à-vis government and its institutions. In one stencil, an image of the sinking Titanic linked the prospects for the Argentine economy to that of the ill-fated ship. The accompanying phrase '*se cayó el sistema*' [the system is down] invoked the computer speak, often heard in Argentina's banks as well as other service outlets where processing equipment routinely failed to operate, leading to long queues of hot, disgruntled customers (Lyle 2007).

Figure 5.8   Disney War Stencil, 2011. BSASSTNCL.
Author's own photograph

BSASSNTCL's most recognised stencil intervention is seen in Figure 5.8. This is the face of former the US President George W Bush, crowned with a pair of Mickey Mouse ears. The original stencil included the slogan 'Disney War'. The mischievous juxtaposition of Bush's portrait and Disney's mouse ears proved engaging across cultures, spaces and contexts. After appearing online, it was quickly viralised, appearing in London, Copenhagen, Costa Rica, Melbourne, Bogota, Tel Aviv as well as other places and contexts far removed from the Argentine crisis. The global popularity of the stencil probably has much to do with its lack of a fixed frame of reference, which allowed it to take multiple 'lines of flight', as Deleuze and Guattari (1987) might put it. Although the stencil projected some familiar words and symbols, these sat together in an ambiguous mixture: 'War' – but which one? Iraq? Afghanistan?; 'Disney' – a proxy for US imperialism perhaps? Or are the Mickey Mouse ears an inference about the competence/ intelligence of the 'world's most powerful man'? Furthermore, redeployment of the Bush/Mouse icon in new contexts across the globe reassigned it with multiple and overlapping meanings, as artists and audiences have brought their own experiences, feelings and ideologies to the interpretive process.

In a similar playful style, one of StencilLand's first interventions, '*El David*', translated Michelangelo's statue of David the giant-slayer into a stencil and placed a kettle and gourd of *yerba mate* – a popular traditional tea drunk in Argentina – in his hands. StencilLand's '*Apocalypsis Love*' meanwhile features a bride and groom in gas masks. The iconography of these stencils has sometimes been linked to national identity and/or social issues in Argentina. Yet, when questioned, StencilLand refutes the idea that there is a fixed political idea behind his work, instead emphasising a deliberate ambivalence:

> behind each of my images is a much darker or twisted story [...] I do not intend to relay a specific message, I do not expect that the viewer 'understands' my ideas exactly; that is not my goal. The main target of my stencils is me. I enjoy the different stages of the process: sketching ideas in a notebook, trialling designs from the PC, cutting the templates and then painting. Commonly this combination is what I enjoy. It may be the case that my work leaves a taste of dissent, but it is rather something internal, perhaps my own self-criticism, that is the 'engine' which brings me back to continue designing over and over again.
>
> (author's translation)

Street art does not just offer an opportunity for play and experimentation however. The experiences of the artists also clearly indicate how engaging in street art production fosters processes of healing and personal restoration, arguments that have a long lineage. While Plato viewed poets and painters with disdain for their acts of imitation, arguing that 'poetry feeds and waters the passions instead of drying them up; she lets them rule, although they ought to be controlled, if mankind are ever to increase in happiness and virtue' (Plato, 360 BCE/1976), Aristotle placed value in the artist's ability to facilitate a purgation of 'pity and fear'. In this sense, the words of the stencil collective, Vomito Attack, are especially interesting.

> we arrived in Buenos Aires in December and the crisis exploded. Without any possibilities of work, no money and a lot of free time – the project started growing. At first we did cut and pastes from newspapers and magazines changing the meaning of the contained information. Then we decided to use the streets as our main canvas, so we translated all that information to stencils and went out to paint.... The name comes from how we recycle images, ideas and information. No fucking copyright exists for us. And also it's a peaceful and good way to get out all the shit that makes us sick.
>
> (Vomito Attack to antrophe 2007)

The collective explain how, jobless and full of energy, they took to the streets to address the kinds of 'shit that made them sick' which by their accounts included rampant consumerism, political corruption, the dictates of the International Financial Institutions and the elitist art establishment. Santiago, one half of the stencilling outfit explains that: 'Making stencils is my way to express my... interior things' (Santiago to Lyle 2007). He explains that they decided to sign their work Vomito Attack because crucially, 'when you vomit it all out, you're better'. Perhaps for this reason, Halsey and Young (2006) argue that street art production does things to the artists' bodies as much as it does things to the surfaces that they inscribe, allowing for processes of healing and renewal. Indeed, many of the stencil makers contend that their art has changed their outlook on politics and the possibilities for change. BSASSTNCL's GG explains how he came to forge new connections and friendships while painting on the streets and realised that 'anyone can express an idea or emotion with just a few pesos'. He claims, 'You don't need the millions that brands pay to be in the spotlight. You don't have to be a politician who pays people to paint his propaganda. You can go out alone and express yourself with a couple of pesos with a can of spraypaint, latex paint and a brush, or a stencil' (GG 2011).

Notably, aided by televisual and web-based transmissions, street art interventions around the Argentine crisis caught the attention of intellectuals, activists and curators from other parts of the world, 'who glimpsed in this turbulent process a novel and vital sociocultural laboratory' (Longoni 2008: 575). Longoni uses the term *turismo piquetero* (picket line tourism) to describe 'the stream of visitors who arrived, armed with cameras and good intentions, to visit neighbourhood meetings, reclaimed factories, pickets and roadblocks'. She explains how in this context, activist art groups, including GAC, *Taller Popular de Serigrafía* and the some of the stencil collectives were subjected to intense attention and wide international circulation. Some were quickly catapulted into the gallery circuit, forming the latest object in the art market's ongoing search for the 'new'.

The practice of stencilling itself became consolidated as a popular artistic genre after 2002. In 2004, the project *Hasta La Victoria, Stencil!* brought together stencil collectives from across Buenos Aires for a stencil 'jam', followed by the publication of a book by the same name. Meanwhile, in 2006, BSASSTNCL participated in a solo exhibition entitled, '*Adentro*' (Inside) at the *Centro Cultural Borges*. Shortly thereafter, a number of collectives participated in the renovation of the dilapidated 'Post Bar' in Palermo, which they decorated in return for rights to use the space as a multi-functional cultural hub, bar, atelier (studio) and taller (workshop) named 'Hollywood in Cambodia'. Post Bar is now one of many stopping points for the capital's popular graffiti tours, managed by enthusiastic expats and self-styled curators, Graffitimundo.

However, tensions are often created when activist art – and street art in particular – is absorbed into the gallery circuit. First and foremost is a philosophical and a practical concern that gallery insertion disempowers the medium of political street art by inverting its quotidian, anti-institutionalist character and stripping it of meaning by placing it in a setting where access is often restricted to an educated, contemplative elite. Indeed, decisions about whether or not to exhibit have sometimes ignited rows among practitioners and led to the break-up of collectives. Vomito Attack is a case in point. Unable to agree on the issue of patronage, the Vomito duo, Santiago and Nico eventually decided to go their separate ways (Lyle 2007). Yet, in spite of initial misgivings, many of Buenos Aires' stencil makers have adapted to the new marketised reality. As StencilLand explains:

> [Painting for money was] something unthinkable and unacceptable to me years ago. I found it very shocking to paint something to sell, It seemed to me that the art it lost its essence, its spirit. When I paint in the street I do not sign my work, do not put my name or my initials, nor my group and much less my email or website. I do not care about fame, I do not pretend to be an artist. But I reached a moment, where the stencils became bigger each time, where I needed to buy more aerosols, where I had to spend more time doing what I wanted: to 'keep painting.' [So I thought] 'Ok. If I sell something, I will invest in aerosols' ... The more I sell the more I can paint... It then turned into an ecosystem in symbiosis. I sell so that I can continue painting on the street.

Argentine street art in the long twentieth century

Drawing on a variety of sources including interviews, archival material and photographs, this chapter offered a somewhat sweeping study of street art from Argentina's Golden Age to the aftermath of the 2001 economic crisis. Covering a range of collectives and interventions, the chapter aimed to illuminate the political work that street art does by moving back and forth between art and activism, strategic and non-rational action. As the chapter shows, in its early stages Argentine street art was shaped in important ways by the import of methods and styles from abroad. In particular, the Mexican muralist David Siqueiros' call to a revolutionary street-based art and his practical guide to stencilling have been heeded and deployed time and again in the Argentine context. However, over the course of the twentieth century, artists and activists have innovated, appropriated and honed an even more expansive range of techniques to galvanise and express *la opina de la calle*. This includes graffiti, stencils, murals, mobile silkscreen, street-based performance as well as more hybrid forms of political street art such as *escrache.*

While the first part of the chapter underlined how street art production became deeply interwoven with the emergence, ascendance and survival of the populist Peronist movement, the second part of the chapter focused more squarely on the role of street art in defiance to authoritarian rule. Influenced by the Situationists in Europe, the *Grupo de Artistas de Vanguardia* sought to converge art with lived reality in Argentina. For them, the streets were the key site and space in which art could be submerged into life and act as a means for subverting and modifying the status quo under Ongania's regime (Carnevale 2016). Both the Experimental Cycle and the Tucumán Arde campaign demonstrated how street poster art and graffiti can be used to reorient the flow and movement of people around the city. It also showed how street art as an alternative or underground medium can broaden public awareness around certain issues and provide an effective source of counter-information in the absence of a free press. Notably, as Argentina moved towards the extreme violence of the Process of National Reorganisation in the mid-1970s, the visibility of oppositional street art receded sharply. McLuhan's maxim that 'the medium is the message' reminds us that shifts in civic modes and forms of expression can tell us a great deal about the underlying conditions of power, and by extension the balance of political opportunities and threats in a given context. The auxiliary street art interventions of the *Madres de la Plaza de Mayo* illustrate how careful, clandestine performances or acts of 'infrapolitics' can be an important strategy for mobilising actors and sowing the seeds of dissent while at the same time shielding activists from the disproportionate violence of the state. Meanwhile, the unprecedented takeover of public spaces as part of the *Siluetazo* signalled the reawakening of civil society and decisive weakening of the regime.

The third part of the chapter moved on to explore street art production during Argentina's democratic period, beginning in 1983. Initially, graffiti and street poster art announced the arrival of new or reorganised political groups and evolved as a means of making demands on the new civilian governments. Key among these have been calls for truth, justice and *reaparición* through novel interventions such as *escraches* and *baldosas* that combine street art with community action to foment collective memory. Lastly, interviews with stencil practitioners from BSASSTNCL, StencilLand and Vomito Attack underline the complex interaction of affect, play and the possibilities for catharsis in street art production, resonating with Julia Tulke's recent work on 'the aesthetics of crisis' in austerity Greece. While it is true that street art manifests in some unique ways during moments of crisis and transition, arguably the medium itself faces a crisis of co-optation in the twenty-first century, as artists and practitioners are increasingly scouted by corporate and state patrons that are keen to capitalise on its allure for tourists and young people. Lewisohn (2008), among others, argues that move from the streets to the galleries in particular, has the tendency to neutralise the political power of street art by robbing it of its context, outsider status and its accessibility. As such, there remains a question as to what comes next for Argentine street art, and whether it can retain its critical edge and power as it is absorbed and accepted into the mainstream of artistic life under global capitalism.

Notes

1   There are other early examples, some of which are of disputed provenance. In *Contra La Pared* (*Against The Wall*), Claudia Kozak (2004) relays the popular myth that in 1840, on his way into exile in Chile, the intellectual Domingo Sarmiento wrote, *'On ne tue point les idées'* (Ideas cannot be killed) along a mountain pass.
2   The Concordancia was a conservative political alliance composed of wealthy agriculturalists and oil interests. Three Presidents belonging to the Concordancia (Agustín Justo, Roberto Ortiz, and Ramón Castillo attained power during "the Infamous Decade" between 1931 and 1943 (Cavarozzi 1992).
3   In the early 1950s the Catholic Church issued underground fliers urging voters to make a choice between 'Christ or Perón' (Chaffee 1993).
4   On the Situationists founding and key ideas, see Matthews (2005).
5   Born in 1942, the artist-activist Graciela Carnevale contributed to the collective projects, Ciclo de Arte Experimental and Tucumán Arde. Keen to preserve the records of these actions she donated documentation and correspondence to ESCALA, where they remain archived. Carnevale was interviewed by the author in 2016.
6   In the nine months between 16 December 1983 and 20 September 1984, the newly established National Commission on the Disappeared (Comisión Nacional sobre la Desaparición de Personas, CONADEP) led an enquiry into the disappearances between 1976 and 1983. CONADEP's full report was issued on 20 September 1984, and was commercially published in a shorter form under the title, *Nunca Más: Informe de la Comision Nacional sobre la Desaparicion de Personas*. It documented approximately 9,000 disappearances, but outlined that due to families' fears of coming forward, the real numbers probably ranged between 10,000 and 30,000.
7   In Nicaragua, the *contras*, backed by the Reagan government, had begun to carry out a systematic campaign of violence, disruption and intimidation. This campaign included attacks on schools, health centres and the majority of the rural population that was sympathetic to the Sandinistas. Widespread murder, rape, and torture were also used as tools to destabilise the government and to 'terrorise' the population into ousting the revolutionary government of the *Frente Sandinista de Liberación Nacional*.

References

Albero, A. and Stimson, B. (1999) *Conceptual Art: A critical anthology*. Massachusetts: MIT Press.

Alston, L. and Gallo, A. (2010) Electoral Fraud, the Rise of Peron and Demise of Checks and Balances in Argentina. *Explorations in Economic History*. 47, pp.179–197.

Betta, L. (2006) *Siqueiros en Argentina*. Blog of Lorena Betta, Professor of Philosophy, University of Rosario. Retrieved from: www.lorenabetta.com.ar/siqueiros-en-argentina

Bishop, C. (2012) *Artificial hells: Participatory art and the politics of spectatorship*. London: Verso Books.

Blanco, P. (2007) Yrigoyen y la Ejercito Nacional 1916-1922. Seminario Taller de Investigación Histórica. Retrieved from: www.madreclarac.com.ar/archivos/Yrigoyen_y_el_ejercito.pdf

Bolton, K. (2014) Perón and Perónism. London: Black House Publishing.

Bosco, F.J. (2001) Place, space, networks, and the sustainability of collective action: the Madres de Plaza de Mayo. *Global Networks*. 1(4), pp.307–329.

Cammack, P. (2000) The resurgence of populism in Latin America. *Bulletin of Latin American Research*. 19(2), pp.149–161.

Canovan, M. (1999). Trust the people! Populism and the two faces of democracy. *Political Studies*. 47(1), pp.2–16.

Carnevale, G. (1968) *Project for an Experimental Art Series*. Originally published as a series of brochurcs accompanying the *Ciclo de Arte Experimental* exhibition of October 1968. Translated by Marguerite Feitlowitz. Available from the *Archivo Grupo de Arte de Vanguardia*, UECLAA Archive, Essex University.

Carnevale, G. (2006) Interviewed for the event 'Agency and Archive at the Royal College of Art'. Transcription by Elena Crippa. Available from the *Archivo Grupo de Arte de Vanguardia,* UECLAA Archive, Essex University.

Carnevale, G. (2016) Interviewed by H.E. Ryan. Feburary 2016.

Castro, A. (2014) Se multiplican en la ciudad las baldosas en honor a vecinos y personalidades. *La Nacion.* Retrieved from: www.lanacion.com.ar/1702220-se-multiplican-en-la-ciudad-las-baldosas-en-honor-a-vecinos-y-personalidades

Cavarozzi, M. (1992) Patterns of Elite Negotiation and Confrontation in Argentina and Chile, *in* Higley, J. and Gunther, R. (eds) *Elites and Democratic Consolidation in Latin America and Southern Europe.* Cambridge: Cambridge University Press, pp.208–236.

Chaffee, L.G. (1993) *Political Protest and Street Art: Popular tools for democratization in Hispanic countries*. Westport, Connecticut: Greenwood Publishing Group.

Collard, M. (2012) The Outing of Torturers in Argentina: Civil Society and the Ongoing Fight Against Impunity. Retrieved from: http://statecrime.org/state-crime-research/the-outing-of-torturers-in-argentina-civil-society-and-the-ongoing-fight-against-impunity/

Deleuze, G. and Guattari, F. (1987) *A Thousand Plateaus.* London: Athlone Press.

Dirks, A. and Siemerink, E. (2013) The Argentina Independent – The Past in the Present: Hidden Places of Memory in Buenos Aires. *The Argentina Independent*. Retrieved from:www.elsesiemerink.nl/2013/03/25/the-argentina-independent-the-past-in-the-present-hidden-places-of-memory-in-buenos-aires/

Druliolle, V. (2011) Remembering and its Places in Postdictatorship Argentina, *in* Lessa, F. and Druliolle, V. (eds) *The Memory of State Terrorism in the Southern Cone: Argentina, Chile, and Uruguay.* New York: Palgrave Macmillan. pp.15–42.

Edwards, T. (2008) *Argentina: A Global Studies Handbook*. Santa Barbara: ABC CLIO

Ehrick, C. (2015) *Radio and the Gendered Soundscape. Women and Broadcasting in Argentina and Uruguay, 1930–1950*. New York: Cambridge University Press.

Favario, E. (1968) Project for the Experimental Art Series, *in* Alberro, A. and Stimson, B. eds (2009) *Institutional Critique: An anthology of artists' writings*. Massachusetts: MIT Press.

Favor, L. (2010) *Eva Perón*. New York: Marshall Cavendish.

Fisher, J. (1989) *Mothers of the Disappeared*. Massachusetts: South End Press.

Foss, C. (2000) Selling a Dictatorship: Propaganda and The Peróns. *History Today.* 50(3), pp.8–14.

Genovese, A. (2001/2013) *Fileteado Porteño*. Retrieved from: www.fileteado.com.ar

Genovese, A. (2007) *Filete Porteño.* Retrieved from: www.folkloretradiciones.com.ar/
    literatura/filete_porteno.pdf

Genovese, A. (2008) *The Book of Filete Porteño.* Buenos Aires: Grupo Ediciones Porteñas.

GG, BSAS Stencil (2011a) Interviewed by Ryan, H. (11 September 2011).

GG, BSAS Stencil (2011b) Interviewed by Escritos en la Calle in Buenos Aires, Argentina.
    Retrieved from: http://graffitimundo.com/interviews/interview-gg-buenos-aires-stencil/

Greeley, R. (2007) Art and Politics in Contemporary Latin America. *Oxford Art Journal.*
    32(1), pp.162–167.

Halsey, M. and Young, A. (2006) Our desires are ungovernable: Writing graffiti in urban
    space. *Theoretical Criminology.* 10(3), pp.275–306.

Hein, H. (1968) Play as an aesthetic concept. *The Journal of Aesthetics and Art Criticism,*
    27(1), pp.67–71.

International Centre for Nonviolent Conflict (2010) *The Mothers of the Disappeared:
    Challenging the Junta in Argentina (1977-1983).* Retrieved from: www.nonviolent-
    conflict.org/the-mothers-of-the-disappeared-challenging-the-junta-in-argentina-
    1977-1983/

International Centre for Transitional Justice (2005) *Accountability in Argentina: 20 Years
    Later, Transitional Justice Maintains Momentum.* Retrieved from: www.ictj.org/sites/
    default/files/ICTJ-Argentina-Accountability-Case-2005-English.pdf

Keck, M. and Sikkink, K. (1998) *Activists Beyond Borders.* Ithaca: Cornell University
    Press.

Kozak, C. (2004) *Contra la pared: sobre graffitis, pintadas y otras intervenciones urbanes.*
    Buenos Aires: Libros del Rojas, Universidad de Buenos Aires.

Kozak, C. (2011) Interviewed by Ryan, H. (8 September 2011).

Latta, M.M. (2002) Seeking fragility's presence: The power of aesthetic play in teaching
    and learning. *Philosophy of Education Archive*, pp.225–233.

Levitsky, S. (2003) *Transforming Labor-Based Parties in Latin America: Argentine
    Peronism in Comparative Perspective.* Cambridge: Cambridge University Press.

Lewis, P. (2006) *Authoritarian Regimes in Latin America: Dictators Despots and Tyrants.*
    Maryland: Rowman and Little.

Lewisohn, C. (2008) *Street Art: The Graffiti Revolution.* New York: Tate Publishing.

Longoni, A. (2006) Is Tucumán Still Burning, Translated by Marta Ines Merajver. Sociedad
    (Buenos Aires). Vol. 1, Selected edition Retrieved from: http://socialsciences.scielo.
    org/scielo.php?pid=S0327- 77122006000100003&script=sci_arttext

Longoni, A. (2007) El Siluetazo (The Silhouette): On the Border between Art and Politics.
    Frontiers, pp.176–186. Retrieved from: www.sarai.net/publications/readers/07-
    frontiers/176-186_longoni.pdf

Longoni, A. (2008) Crossroads for Activist Art in Argentina. *Third Text.* 22(5),
    pp.575–587.

Longoni, A. and Bruzzone, G. (2008) *El Siluetazo.* Buenos Aires: Adriana Hidalgo.

Lyle, E. (2007) Shadows in the Streets: The Stencil Art of the New Argentina, *in* MacPhee,
    J. and Reuland, E. (eds) *Realizing the Impossible: Art Against Authority.* Edinburgh:
    AK Press.

Marx, G. (1990) Argentina's President Pardons Leaders Of 'Dirty War' On Leftists.
    Chicago Tribune. Retrieved from: http://articles.chicagotribune.com/1990-12-30/
    news/9004170832_1_pardoned-two-former-military-presidents-dirty-war

Matthews, J. (2005) *An Introduction to the Situationsists.* Anarchist Library.

Museo de Arte Contemporáneo de Rosario (2009) Siluetazo. Retrieved from: www.
    macromuseo.org.ar/coleccion/artista/e/el_siluetazo.html

Nouzeilles, G. and Montaldo, G. (2002) Splendor and the Fin de Siecle, *in* Nouzeilles, G. and Montaldo, G. (eds) *The Argentina Reader: History, Culture, Politics.* Durham: Duke University Press.

Padin, C. (1997) *Art and People*. English Online Edition. Retrieved from: www.concentric. net/~lndb/padin/lcptuc.htm

Pisani, A. and Jemio, A. (2012). Building the Testimonial Archive of the Operativo Independencia and the Military Dictatorship in Famaillá (Tucumán, Argentina): A Critical Review. *Oral History Forum.* Retrieved from: www.oralhistoryforum.ca/ index.php/ohf/article/view/450/521

Romero, L. (2002) *A History of Argentina in the Twentieth Century.* Pennsylvania: Pennsylvania State University Press.

Seidman, S.A. (2008) *Posters, Propaganda, and Persuasion in Election Campaigns Around the World and Through History*. New York: Peter Lang.

Sitrin, M. (2012). Horizontalidad and territory in the Occupy Movements. *Tikkun.* 27(2), pp.32–63.

Stein, P (1994) *Siqueiros, His Life and Works*. New York: International Publishers Company.

StencilLand (2011) Interview with Holly Ryan, *in* Ryan, H. E. (2015) Affect's Effects: Considering Art-activism and the 2001 Crisis in Argentina. *Social Movement Studies,* 14(1), pp.42–57. September, Buenos Aires.

The Economist (2015) *The persistence of Peronism*. Retrieved from:www.economist.com/ news/americas/21674783-argentinas-dominant-political-brand-defined-power-not-ideology-persistence

Upton, B. (2015) *The Aesthetic of Play*. Cambridge, MA: MIT Press.

Vallejos, S. (2008) El regreso de las super amigas. *Pagina 12.* Retrieved from: https:// translate.google.co.uk/translate?hl=en&sl=es&u=http://www.pagina12.com.ar/diario/ suplementos/las12/13-4511-2008-11-21.html&prev=search

Veigel, K. (2009) *Dictatorship, Democracy, and Globalization: Argentina and the Cost of Paralysis, 1973–2001*. Pennsylvania: Pennsylvania State University Press.

Veroni, R. (n.d.) 'I've always been interested in popular forms of expression' – Interviewed by Escritos en la Calle. Retrieved from: www.escritosenlacalle.com/pdf/graffiti/ I've%20always%20been%20interested%20in%20popular%20forms%20of%20 expression%20-%20Interview%20with%20Ral%20Veroni.pdf

Vezzetti, H. (2009) Argentina: The Signs and Images of 'Revolutionary War'. *German Historical Institute Bulletin.* 6, pp.27–31.

Vomito Attack (2007) Interviewed by Antrophe. Retrieved from: http://soundtracksforthem. blogspot.co.uk/2007_08_01_archive.html Accessed: 12 December 2012.

Whitener, B. (2009). *Genocide in the Neighborhood*. Oakland: Chainlinks.

6 Conclusion

This book offered up the term 'political street art' as a loose category for interventions whose creative and material use of the street is in some way tied to their political meaning. It held that to be *political* is not just to express political opinions but rather to be oriented towards society, to engage with and challenge existing structures and terrains of power. Following Ranciere, it argued that politics by definition involves a 'blurring of boundaries'; the folding of new issues, areas and practices into the political field. The cases explored in this book illuminate the variety of political work that has been done by street art in Latin America. From indigenous peoples in Bolivia and anti-dictatorship activists in Brazil to the mothers of the disappeared in Argentina, street art has offered voice and prominence to groups that have been 'excommunicated' from political processes by raced, gendered and socio-economic hierarchies, repression and fear. However, as this book has sought to illustrate, producing and engaging with political street art is a complex process, and one which mainstream social movement theory in its strong structuralist and rationalist modes, fails to fully capture. As Chapter 2 argues, when political street art is viewed through the lens of political process theory in particular, it begins to look functionally indifferent from other 'contentious performances', conceived of as just another instrument for strategically framing issues. The task of this book has been to look more closely at political street art in order to discern its dynamics and nuances, taking cues from approaches grounded in a practical or heteronomous aesthetics.

As Charles Tilly (2008: 11) put it, 'no two contentious performances mirror each other perfectly. Indeed they would lose some of their effect if they operated like a military drill'. Hence, the styles, forms and themes taken up by street art practitioners in Brazil, Bolivia and Argentina have varied according to national political events, cultures and experiences. The *Grupo Tupinãodá* collective drew on an intellectual legacy of *anthropophagy* or intellectual cannibalism, particular to Brazil. Street art interventions have played an important, yet underexamined, role in amplifying indigenous demands and negotiating indigenous identities in Bolivia. Meanwhile in Argentina, the evolution of political street art cannot be considered apart from the rise and resolve of the Peronist movement and the birth of mass politics. However, the combination of unstructured interviews, observation and archival material gathered from fieldwork in these three Latin American nations does make it possible to draw out some common threads, lessons and characteristics that enhance our understanding of street art and the political work that it does.

As the previous chapters demonstrate, from (at least) the early twentieth century, street art has been mobilised time and again as an instrument of protest and means of expression in Latin America. Emerging as a low technology means of mass communication, it has served an important function in circulating political messages, both pro-system and anti-system, among groups lacking access to other information channels. During the years of dictatorial excess which assailed populations in all three countries under study, street art interventions became an important source or example of 'infrapolitics'. Street graphics, produced stealthily, would announce protest marches or strikes, project the logos, colours and claims of opposition groups and disrupt the authoritarian aesthetic with free forms of expression and play.

Repeatedly, at times of both authoritarian and democratic government, street art has turned the urban surfaces and streets of cities such as Sao Paulo, La Paz and Buenos Aires into communicative media with the potential to instruct, educate and mobilise. Yet, examples such as painting, ad-jamming, and street theatre are far from passive transmission points. They may also be performative in as far as they mediate within, challenge and even alter the political status quo. Argentine *escraches* for example, do not just call for an end to impunity, they enact accountability through a public shaming ritual. Painting too can disrupt and reconfigure public spaces, playing with the urban aesthetic to probe at existing boundaries and construct 'anti-environments' through which new possibilities for thought and action may be revealed. In this way, thinking about street art prompts a re-examination and extension of the concept of 'political opportunity' as it has been deployed in mainstream social movement studies.

A fundamental part of street art's political power lies in its ability to tap into and transfer latent sensations, sentiments or 'affect'. Indeed for practical aestheticians like Jill Bennett, affect is the natural medium of aesthetics. Feigenbaum, McCurdy and Frenzel (2013) distil existing work on affect in social movements into three useful categories, describing it as *pre-discursive, communicable and potentially transformative.* Firstly then, affect describes sensations or felt intensities that we don't (yet) have the right words for. Secondly, these sensations may travel or transfer from one object or body to another, with media forms acting as amplifiers that extend their reach and resonance. Thirdly, in as much as such sensations effect what the body can do, they can be transformative, foregrounding new ways of understanding oneself in relation to the world around. Each of these categories is helpful for understanding the encounters seen in and around political street art. For example, street art production and reception sometimes connects with the feelings of practitioners, activists and audiences in ways that words and text cannot. This observation is particularly useful for understanding and engaging with street art outpourings such as the '*Siluetazo*' and the stencil movement that accompanied the 2001 crisis in Argentina. Here, affective states prompted people to act – and interact – in new ways, taking up street art production as a way of working through a shocking or disquieting episode. The notion of a therapeutic or cathartic effect in art production has a significant pre-history, finding its first articulations in the work of Aristotle, who placed value in the artist's ability to facilitate a purgation of negative feelings. Fast forward by a couple of millennia and we can see some similar arguments today in the work of psychologists and trauma therapists who have referred to creativity as a possible 'resilience factor' (Phillips 2012) and 'conduit for undigested [or indigestible] material' (ibid.).

In recent years, we have seen some street art 'outpourings' carry across nations, cultures and borders, assisted by new and increasingly accessible mobile, digital technologies. The sharing of images through digital networks and social media platforms has made street art increasingly visible across the world and street artists themselves are acutely aware of this. Many have deliberately shifted their practice to address a more global constituency and/or contribute to a continually growing digital archive of works by documenting and uploading photographs of each new intervention to Twitter, Flickr, Instagram or other platform. This online sharing of street art across the globe has given birth to an almost paradoxical set of processes, whereby street art styles and motifs have tended to converge, while simultaneously being reinterpreted by increasingly diverse audiences, each of which brings its own culture, experiences and machinations to the decoding process. Though closely tied to locations, cultures and the temporal performative act of making, the practices of street art, as well as the works themselves today 'vacillate between the specific materiality of urban space, street locations, local contexts, and the exhibition, distribution, and communication platform of the Internet and Web' (Irvine 2012: 10). On the surface then, it seems that street art at once reflects and resists the assimilative tendencies associated with globalisation. However, to the extent that the worldwide web becomes a repository for an ostensibly 'global street art', it takes on one of the functions traditionally reserved for the gallery: that is, freezing and anchoring an image in time and space. If street art derives at least some of its political potency from its ephemeral form, then this is a development that surely warrants pause and exploration. While the political work of street art in an age of digital reproduction goes well beyond the scope of this book, it is hoped that the material contained herein will serve as a useful cue for further discussion and research in this evolving area.

References

Feigenbaum, A., McCurdy, P. and Frenzel, F. (2013) Towards a method for studying affect in (micro) politics: The campfire chats project and the occupy movement. *Parallax*. 19(2), pp.21–37.

Irvine, M. (2012) The work on the street: Street art and visual culture , in Sandywell, B. and Heywood, I. (eds) *The Handbook of Visual Culture*. London and New York: Berg. pp.235–278.

Phillips, A. (2012) Art and Trauma: Creativity as a Resiliency Factor. Good Therapy. Available from: www.goodtherapy.org/blog/art-and-trauma-creativity-as-a-resiliency-factor-0307124

Tilly, C. (2008) *Contentious Performances*. Cambridge: Cambridge University Press.

Index

Page numbers in *italics* refer to illustrations (e.g. *94*). Information in notes is indicated by the page number followed by n and the note number (e.g. 12n1).

148 *Index*